ROOTS OF DEFEAT

ROOTS OF DEFEAT

A YOUNG CAPTAIN'S EXPERIENCE OF THE VIETNAM WAR, 1955–1963, AND WHAT CAME AFTER

A memoir of the war and what went wrong

MICHAEL J. FRANKWICZ

Published by Boodles Press, Hendersonville, North Carolina

Library of Congress Control Number: 2025926881

Paperback ISBN: 979-8-9934356-0-2
Hardcover ISBN: 979-8-9934356-1-9

Cover map: "Vietnam and Neighboring Countries in 1968." Wikimedia
    Commons. Public Domain.
Book cover and interior design by Rebecca LeGates

First edition 2026

CONTENTS

FOREWORD

REBECCA J. FRANKWICZ

MY FATHER, MICHAEL J. FRANKWICZ, was born in 1935 in Milwaukee, Wisconsin, and grew up an hour away, in Janesville. He attended the University of Wisconsin into his junior year but dropped out for financial reasons and then enlisted in the army in 1955, when he was twenty years old. After basic training at Ft. Leonard Wood, Missouri, he went to Ft. Dix, New Jersey, for advanced infantry training; then to Ft. Benning, Georgia, to the Infantry Officer Candidate School; and finally on to jumpmaster school, where he became a second lieutenant of paratroopers. From there, he was ordered to Schweinfurt, Germany, to the 86th Infantry of the 10th Mountain Division. He spent three years on duty in Germany before being assigned to St. Louis, and it was there that he met my mother, Peggy Poole, whom he married in July 1961. In early 1962, now a captain, my father reported to the Army Language School in Monterey, California, to study the Vietnamese language. He shipped out to Vietnam in May 1962, and my mother returned to St. Louis, her hometown.

When my father left the military in 1964, he worked for the St. Louis County Civil Service Commission while taking night classes at Washington University, after which he started a career in the computer industry. My parents had seven children (I'm number six), and they raised us in St. Louis, where my father ran his own business and was active in his church community. He started writing this book in the mid-1990s, which is around the time he was diagnosed with colon cancer. He passed away on January 21, 2001, when he was sixty-five.

I moved to North Carolina for college and never returned to live in St. Louis. My dad and I didn't connect as much as I now wish we had, especially to talk about his book and his time in Vietnam, but I also realize that I would not have had the perspective I needed to really grasp his military history. Not to mention that there was more than one version of his experience in Vietnam: one he would share with his family (although he didn't talk about Vietnam much), another that the reader will learn in this book, and, I believe, still another version beyond that—one so soul-crushing that it remained locked inside. Until I started working on his manuscript, I had no clue what my dad went through during the war and what he continued to experience after he came home.

I knew my father to be a warm and generous man who did the best he could with what he had to work with. If I had to describe him in just a few words, I'd say that he had a huge heart and was committed to making life better for people. Even while he was fighting colon cancer, he was going into the inner city to feed the hungry and share his faith. He wanted people to have what they needed, and he wanted to save as many souls as he could before he died. That big-heartedness speaks to who he has always been. In the pages that follow, you can see this trait in how he stood up for his fellow soldiers and how he connected with the people of Vietnam, especially those in the most unfortunate circumstances.

I see now that Dad had to be two people in his life, the warrior and the civilian. He was hurt by how he was treated by both the government/military and many antiwar civilians after he left the service, and this book was one way he worked through those feelings. As he read the books that were being published in the mid-1990s, after the declassification of important documents about the Vietnam War, the process clearly opened some old psychological wounds. He spent a lot of time and energy working on this manuscript, compiling his old letters and articles from the 1960s and writing new annotations and essays during the 1990s. I know he hoped one day to publish the result for others to read.

One thing I found notable while reading through this material is that most of his frustrations were not just about his own experience but about how other people in his position were similarly mistreated. He complained up the chain of command to the people who had real power to change things, but unfortunately they were sometimes the same people he felt were playing too casually with the lives and futures of others. It's interesting to see that, in many cases, he placed the blame not on specific individuals but on corrupt systems. Over the years since my father passed away, so much has come to light that only validates his thoughts about the military and government that sent him to Vietnam.

On Veterans Day of 2024 I decided to honor my father by taking what existed of his work—one three-ring binder of typed pages and a floppy disk(!)—and put my own efforts into turning it into a published book. The original letters he wrote home from Vietnam give us a picture of what kind of person he was then and how his earlier life shaped his experiences in Vietnam, and the later pieces reflect his ideas as a man with a deeper understanding of what happened. His postwar thinking was in passionate reaction to Robert McNamara's books *In Retrospect: The Tragedy and Lessons of Vietnam* (1995) and *Argument Without End: In Search of Answers to the Vietnam Tragedy* (1999), while on the other hand he largely praised Fletcher Prouty's *JFK: The CIA, Vietnam and the Plot to Assassinate John F. Kennedy* (1992), Paul Hendrickson's *The Living and the Dead: Robert McNamara and Five Lives of a Lost War* (1996), and Francis X. Winters's *The Year of the Hare: America in Vietnam, January 25, 1963–February 15, 1964* (1997). (Other sources he used can be found at the back of this book.) Some of his later writing is not included here, either because it was unfinished and/or highly technical or because now, almost thirty years after he wrote about them, the major players have passed away and the systems, structures, equipment, and political atmosphere have changed so drastically.

That said, this version of my father's book focuses on what will always be relevant: one man's firsthand experience of the war and its aftermath. As the story of a soldier who went in fully committed and came home disillusioned, it is an important addition to Vietnam War literature. But in the end, his perspective is that of one person, and I hope it will find readers who connect with it on that level. This isn't just a book about why the war went wrong but about how to stand up for what you believe in, for what's right, even when everyone around you is telling you to give in and get along, whether to promote yourself or just to protect your own interests.

In his writing, I see my dad's resiliency and his amazing ability to adapt and survive while staying true to his core values. This is how he lived his whole life. I hope you find something valuable in what he has to say.

Rebecca J. Frankwicz
2026

INTRODUCTION

MICHAEL J. FRANKWICZ

THIS PROJECT BEGAN when one of my children prodded me to write down my Vietnam wartime experiences for the sake of our descendants. From the Civil War on, no one in my family had done so, and something was lost that shouldn't have been.

As the reader will see, I have long been a student of Southeast Asia. Beginning in the late 1950s, I studied to prepare for my tour there, and I picked up my research again in 1994. I decided I had something to say to a wider audience, and that is how this book came to be. In the pages that follow, I make note of the many sources I referenced while writing about the war, and there is also a list of these sources in the back of the book. To the reader who wants to know more about what was happening in Southeast Asia prior to the war, I recommend L. Fletcher Prouty's *JFK: The CIA, Vietnam and the Plot to Assassinate John F. Kennedy* (1992). For the most part, I agree with his re-creation of the history and his takes on what happened and why, and the book will give you a good understanding of what preceded my experiences in the war.

I must admit that one of my main motivations to compile my thoughts about the war into a book was the anger I felt after reading Robert S. McNamara's *In Retrospect: The Tragedy and Lessons of Vietnam* when it came out in 1995. Then, when his *Argument Without End: In Search of Answers to the Vietnam Tragedy* appeared in 1999, this project was near its end, but I felt a need to rebut many issues covered by Mr.

McNamara (secretary of defense under Kennedy and Johnson) and his co-authors. [Editor's note: The version of the book you're reading focuses more on the author's personal experiences in the war and less on rebutting McNamara and "setting the record straight" about what happened in Vietnam and who might have had the information and wisdom to change the course of the war and save many lives. The next two paragraphs provide a condensed overview of the author's problems with McNamara's take.]

Argument is loaded with all kinds of wisdom in 1999 that was nowhere evident in 1961–63. When the advisors in the field at the time tried to offer feedback to the top leadership, we were ignored, and if we persisted, we tended to get into trouble. In fact, it finally came down to being forced by the situation to resign and destroy your career if you couldn't live with the way it was going to have to be. (This is what happened to me.) There was also a lot of Communist rhetoric and propaganda in *Argument*, and it does a good job of telling the Communist side of the story. It does a poor job of presenting *our* side, so it leaves a very distorted version of history, maybe without attempting to. The basis of what the Kennedy administration did in Vietnam was to a large degree founded on our belief in the legitimacy of our position (that is, South Vietnam's position) regarding the Geneva Accords of 1954. My US Army comrades and I were personally sent into the fight on that basis. Some of us were killed in action, some of us were maimed, some of us had our careers destroyed when we appealed policy over these matters, and some of us became scapegoats in the society we returned to, while the civilian elitists who were orchestrating the war escaped unscathed.

As a former officer of the United States Army committed into combat by Mr. McNamara and his people, and as someone who remembers the names of my comrades on the first panel of the Vietnam Veterans Memorial Wall in Washington, DC, I cannot accept much of what is in *Argument*. On The Wall, the first panel to the right of center, one of the two largest, contains the names of those killed in action (KIA) from 1961 to 1963. The rest of The Wall contains what I believe to be the fifty-eight thousand unnecessary KIA that came in the years after. I don't think our government performed with due diligence when they put us there, and I think this McNamara team is once again showing elitist arrogance in conducting the inquiry in *Argument*. There was and is great disrespect shown to the troops. Many of us were scapegoats in the halls and streets of America for what they put us into. But we have the right to our opinions as well, and this book aims to share this one man's opinion.

In addition to the "War Diaries" part of this book (chapters 2 through 6, the bulk of which consists of excerpts from letters I wrote home to my wife and mother), I have included pieces I wrote in the years immediately following the war (chapter 7) and pieces I wrote in the late 1990s specifically for this book. These writings contain historical context, information about my personal background, essays on military theory, proposals for tactical operations, and rebuttals of some of the literature that has been published in the wake of the war.

One of my main goals in writing this book has been to prove how many of the mistakes could have been avoided had the people in charge bothered to listen to those of us who had studied guerilla warfare, made an effort to understand the South Vietnamese people, and had real firsthand experience of what it was like on the ground in the early years of US involvement in the war. I am highly critical of how the Kennedy administration handled Vietnam. It seemed to me at the time that the operations of the North Vietnamese more or less followed patterns of Communist takeovers that I had studied, and it was very discomforting to be right there when it happened, to know and understand the significance, and to watch our own leadership get it wrong again and again.

I believe that the war could have been winnable at any time *if* the United States had been motivated to do so, but I believe that there was one particular small window of opportunity between late 1961 and early 1963—my time in the war—when we had the chance to launch a winning policy *at a reasonable cost* that the American people would have accepted. When that failed to take place, I regret to say that nothing else counted because America was not willing to pay the vastly higher price later. So much tragedy and loss could have been prevented. Here is the story of those years from my perspective.

PRE-VIETNAM

CHAPTER ONE

BACKGROUND OF A YOUNG WARRIOR

1955–1960

I WAS WELL INTO my junior year at the University of Wisconsin before I ran out of money and was forced to drop out. I decided to enlist in the army before I continued on with college. So, in September 1955, I reported to Ft. Leonard Wood, Missouri, for basic training in the 86th Recon Battalion of the 6th Armored Division. I quickly found out that my National Guard and ROTC experience could pay off. They made me a trainee platoon sergeant, and I took the Officer Candidate Test, scoring 150, said to be the third-highest score at Ft. Leonard Wood to date. I felt pretty good about that, since I was up all night before on duty. From Ft. Leonard Wood I went to Ft. Dix, New Jersey, to the 69th Infantry Division for advanced infantry training. Again I served as a trainee NCO (non-commissioned officer) and then as an instructor in field fortifications while I awaited the OCS (Officer Candidate School) board['s orders].

One cold day in New Jersey we were running a squad attack exercise in which we had to cross a frozen swamp and storm up what we thought at the time was the only hill in New Jersey. They threw some training explosives to simulate incoming mortar fire, so we all belly-flopped on the swamp and broke through. Delighted, we picked ourselves up and floundered into a ditch at the base of the hill. As the squad leader, I fired a red smoke grenade from my M1 rifle and led the charge up the hill. Since I was still pretty green at this, it was heady stuff, and it wasn't till I got pretty well up the hill that I discovered my rifle was broken in two, held together by just the strap. I must have stuck the muzzle into the mud when I'd broken through the ice in the swamp. When I fired the smoke, the M1 blew up in my face and I didn't even know it. I was very fortunate not to have been seriously wounded. Someone was looking out for me.

After some time, they finally put me on a train to Ft. Benning, Georgia, to the Infantry OCS. About 120 started out and 59 graduated. I believe I graduated about thirty-ninth. I formed friendships there that lasted throughout my ten years in the army, as we kept meeting up with various classmates here and there. From OCS I went through jump school and jumpmaster school, becoming a second lieutenant of paratroopers.

Graduation from jump school was a memorable formation. It was late afternoon on a wet fall day, with the sun peeking through the glistening, wet foliage. We gathered in a battalion-sized group up at the post chapel in Ft. Benning, and we were all decked out in ODs and pinks and greens with bloused boots and airborne paraphernalia. (The army was just about to switch uniform styles from the World War II–era "pinks and greens" and ODs [olive drab green uniforms?] with brown shoes to the new "army green" with black boots.) We carried M1 rifles with fixed bayonets. Several of the army's legendary paratroopers from World War II and Korea presided, and the battle flags with combat streamers drifting in the wind made the whole scene a most stirring military event. This wasn't just a parade; this was a formation with some of the army's elite.

From Ft. Benning, a large number of my OCS classmates and I were ordered to Schweinfurt, Germany, to the 86th Infantry of the 10th Mountain Division, leaving on a troopship from the Brooklyn Navy Yard shortly after Thanksgiving. A huge aircraft carrier was in the harbor as we pulled out, an impressive sight. There were about three thousand troops and dependents on board, and myself and another second lieutenant were each given command of half the troops for the trip, both of us reporting to a field grade officer. That got us seated at the captain's table in the mess.

About three days out we got into a violent storm, which one navy chief with gold stripes told me was the roughest of his career. The ship was pitching violently, and I had to patrol all over the ship, outside and inside, and I was thoroughly awed by the sea. Earlier, there had been problems between some of the female dependents and some of the troops, and as a result, we'd had to punish some of the men by putting them in the brig, which was located in the bottom of the bow—not a nice place to be in such a storm. (Some of the wives had been sent home on a different ship.) There was a large latrine up in the stem, and I remember hanging on in there during my patrols, looking out the back door and watching soldiers hanging over the railing, vomiting into the sea, as the stem of the ship seemed to rise and point to the sky and then tumble down into the deeps. I didn't get seasick until the last day of the storm, so I didn't suffer much. For my service on the trip, I received a letter of commendation for being a troopship commander, and I never could figure out if I'd really done that good of a job or if the field grade officer liked me or if the commendation was just a custom.

Finally, we passed the White Cliffs of Dover and soon came into Bremerhaven Harbor. My first impression of Germany was two huge leather-coated German policemen who came on board and proceeded to the galley for a snack.

At Bremerhaven, we ran a covered tunnel ramp from the ship down to a troop train, and military policemen were spotted all around as we downloaded the troops into the train. Another second lieutenant and I were made troop train commanders, with the mission of delivering the train to Stuttgart. We were put in a nice compartment in the last car, while the troops were all packed in like sardines in a long string of cars, with two or three mess cars in the middle. At mealtimes, the troops from the rear cars would be sent forward to pick up their food in the middle cars, and then they'd walk all the way through to the front of the train, where they'd stand or squat in the aisles to eat their food. When finished, they would file back through the cars, washing their dishes in the middle before heading on back to their bunks in the rear. Then, the front half of the train would do the same thing the other way.

When we got to Bremen, the train stopped and I jumped off to chat with some British RMPs. Suddenly the train started chugging off and I had to run to get on. Angry, I grabbed the German conductor and snapped, "You *do not* move off without the troop train commander!" By this point, I was pretty over-impressed with myself. "Herr Leutnant," responded the unimpressed conductor, "in Deutschland the trains move on schedule regardless of who is not on board." So I learned a lesson, as second lieutenants have to do.

In Stuttgart, we signed over the troops and I soon found myself in a first-class compartment on another train heading for Würzburg. I had only one companion, a middle-aged American lady who was married to a wealthy Argentine. The scenery was fascinating and all very new to me, and the lady was an interesting companion, as she could explain the countryside and European visitor tips in general, and she was very interested in hearing the latest home news. At Würzburg, I changed trains and road coach class, crammed with all manner of Germans, none paying me much attention or showing any inclination to speak English. I was getting a kind of lonely feeling. When I arrived at Schweinfurt, a soldier driving a three-quarter-ton weapons carrier was waiting for me and carted me off to the officers' club at a former Luftwaffe base at the edge of town. Renamed Conn Barracks by the Americans, the place was palatial and deserted, as the troops were in the field.

I settled in, was issued my field gear, and then out to the field I went. I was given a rifle platoon in B Company, 86th Infantry. Among my first memories there was being isolated with the platoon on a snow-covered, sparsely wooded hillside in a defensive position. This was at the tail end of the Hungarian Revolution of 1956, and we could hear in the distance the revving up of diesel engines of what we were told were Russian T-34/85 tanks. Things were touchy near the border.

The B Company commander was a fine-looking older captain who was well decorated from the Korean War. I ran the rifle platoon for a while and quickly found myself handling most of the company's training, as all the other platoon leaders were on special duty elsewhere. The majority of the time I was the only officer the company had with them.

Soon I was transferred to take the weapons platoon, which gave me heavy weapons experience. I remember hauling our recoilless rifles up a rocky cliff one winter day to get a good field of fire. In those days the big maneuvers and field training took place in the wintertime, so we went on a big maneuver in February 1957. We started somewhere northeast of Schweinfurt, up near the Fulda Gap, I believe, and we fell back to successive positions until we ended up on the south bank of the Main River near Aschaffenburg. The maneuver enemy was a battalion from one of the other regiments, and they were giving us a pasting. There was plenty of cold and snow and rain. At one point, Captain W——came back with an order from battalion and gathered the officers and key NCOs together. The order he gave me was to take a large chunk of B Company to a certain location and be prepared to execute a specific operation in coordination with other elements of the battalion. I did what he said, but when we arrived we were obviously in the wrong place for the battalion's requirements. The officers and NCOs with me agreed that we had done what the captain said; we were all there together when he'd given the order. The captain and the rest of the company, however, were nowhere to be seen.

Things fell apart then (including the weather), and the 86th crumbled backward in disarray. At about 3 a.m., in a rainstorm and with wet, tired, and hungry troops, we arrived at a hillside on the south bank of the Main, at our last position, and there stood the captain and the battalion commander trying to sort out the vehicles assembling there. Myself and a couple of senior NCOs who had been at the op order stumbled out and approached the two commanders, saluting. "There he is!" shouted Captain W——. "There is Lt. Frankwicz, who screwed it all up!" The two sergeants and I looked at each other, aghast, as we realized something had gotten fouled up and I was getting the blame for it. It turned out Captain W——had sent us to the wrong place, panicked, and then lied, blaming the screwup on me to save his own tail. B Company knew it, and this became the main gossip topic among the officers and men for weeks. A few months later the battalion commander was cashiered out and came back in as a corporal. Captain W——survived that one.

In late winter and early spring we went back out on maneuvers, this time acting as the "aggressor" against the 85th Infantry out of Bamberg. The 1st Battalion simulated a much larger force, overrunning position after position of the 85th, even worse than we got it. The fun of it was that we officers of B Company had free latitude to plan our own operations, and we were doing so with ever-increasing success. The troops quickly caught the fever of success and behaved with skill and spirit. B Company really came together in that operation, and even Captain W——got caught up in it.

One stirring memory from that European service was the evening ritual at 5 p.m., when a bugle would blow "To the Colors" and bring down the flag. All traffic stopped, all military personnel saluted or presented arms. A cannon fired as Old Glory came down to the Regimental Color Guard. It was the kind of moment, done there in front of the Germans, that made us proud to be American soldiers.

That said, not everything was going smoothly. One of the problems was that the TO&E [Table of Organization and Equipment] was apparently not working well. The old triangular structure wasn't suited to the kind of fluid situation our manner of combat was then requiring. We also had a problem with some of our leaders not being very good. Certainly I had a great deal more respect for most of the middle-management leaders, such as those I served with in Augsburg in '59. In any event, we spent much of the rest of my time in Schweinfurt converting to the Pentomic TO&E put in by General Maxwell D. Taylor. Under the new structure, the 86th Infantry Regiment, a unit designation not renowned for a long and glorious history, was replaced by two battle groups made up of the same personnel: the 2nd Battle Group 7th and 10th Infantries. The 7th, known as the Cottonbalers, had sixty-two battle streamers hanging from its colors, the most of any.[1] It had a much-respected recent history as part of the 3rd Infantry Division in World War II and Korea. Being a student of military history and seeing how unit tradition worked in the British Army and in my own short stint with our paratroopers, I was keen on building on the traditions of the 7th.

The Pentomic structure also added to the 86th a fourth rifle platoon, and it up-gunned heavy weapons to the rifle company, including mortars from 60mm to 81mm and recoilless rifles from 57mm to 106mm. The battle group was commanded by a full bird colonel. No more regiment or brigade, no more battalions for lieutenant colonels. The result was great for young company grade officers like myself, who were actually permitted to handle the troops rather than be sent off on all the special duty. I now ran a task force–sized unit that you could call a baby battalion, and I was deployed with a lot of the independence and initiative expected of the commanders. The new structure was, however, a tragedy for field

---

1. Each battle group had "colors," which were large flags that carried the unit's earned battle streamers (colored ribbons with battle information that tell the unit's history). The 7th Infantry had more streamers than any battle group in the army at that time. These were emblems of a unit's pride. In the 7th, we also had a big cotton bale sitting on our parade ground, because the 7th was nicknamed the Cottonbalers after holding positions behind cotton bales during the Battle of New Orleans in 1815. One time, some pranksters from our sister battle group, the 10th Infantry, thought we were overdoing the tradition stuff, and one night they sneaked out and set fire to our cotton bale!

grade officers (majors, lieutenant colonels, colonels), many of whom had been young company grades (first and second lieutenants or captains) in the Korean War. Now, the few who would survive to full colonel would have to wait till then to get a line command. As I was to find out later, many of the officers who had staff jobs at higher command levels took to interfering with and micromanaging the companies.

One interesting aspect of the Pentomic concept that carried on into later years in the army was that we became combined arms officers instead of just infantry officers. We worked more with armor, functioning often as mechanized infantry, and we worked more with helicopters developing future air assault tactics. We were also trained in the tactical deployment of nuclear weapons, if necessary. I loved working with armor and was an amateur student and admirer of the blitzkrieg and Patton tactics, but I learned most of my handling OTJ (on the job) and made some mistakes on the way, such as the day I ran a company attack exercise and sent a flanking platoon mounted in armored personnel carriers into a swamp, sinking, sinking, sinking . . . oh, trouble.

Even as a young army officer, I had a strong motivation to succeed. Since boyhood days I had been a student of military history, and as I read voraciously I also kept copious notes about various military topics. As a result, by the time I was commissioned in the infantry, I had an unusual appreciation of not just the other services but also the other branches of the army. I was able to work on a combined arms basis without the tunnel vision of only having my infantry training.

I was also interested in the study of guerilla warfare and counterguerilla warfare. As an infantry unit commander in Germany during the late fifties, we, as part of NATO, were confronted with a Soviet opponent who heavily outnumbered us. As the Red juggernaut rolled westward, there was a possibility during the early days of fighting that some of us could be cut off and badly mauled. In preparation, we conducted training in escape and evasion and in behind-the-lines resistance, or guerilla warfare. Since I was the officer who conducted the training, I researched the subject to learn all I could.

Becoming intrigued with this aspect of warfare, I accumulated a file and a library of every English-language and translated book that I could find on the subject. Since I knew that this was a different kind of warfare than our normal training provided for, I took pains to prepare myself, something that, frankly, few did, or at least none that I knew. I was a thoughtful and studious soldier who wanted to think through the military problems confronting us. I was not deterred by the fact that I was not a West Pointer or that the powers that be had a "who are you?" attitude. On the other hand, I believe I had a reputation amongst my peers for being a competent, dedicated soldier. My best friends were that kind of soldier too, and our friendships were based upon mutual respect. My studies in military history convinced me that the army needed thinkers like us, and I believed that, somehow, eventually the army would reward its thinkers. What I was about to find out was that the top levels in government and in the bureaucracy of the army were about to embark upon an era of ignoring or destroying anyone like me who spoke out and didn't knuckle under. Independent initiative or thinking outside the system could mean trouble.

In 1958 I was performing staff duty at Headquarters, Seventh Army, in Stuttgart-Vaihingen. That was a forty-hour-a-week job, and I saw much of Europe that year, including the 1958 Brussels World's Fair. It was during this time that I became acquainted with General Harold K. Johnson, then chief of staff of Seventh Army and later chief of staff of the army during the Vietnam conflict. Wanting to do something worthwhile with our time, I, together with one of my buddies, Bob Hatcher, a Chemical Corps officer on staff at Headquarters, Seventh Army, took over the Boy Scout troop for the army dependent kids there. Through that activity, I became acquainted with General Johnson, who used to come around with his son Bobby. My friend Bob and I ran that Boy Scout troop with all our hearts, just like we soldiered, and I think General Johnson noticed that. He was a fine man, possibly helped my career later, but he and the other top generals during the Vietnam War made, I think, a big mistake by not speaking out against the way things were happening.[2]

In 1959 it was back to the line, to the 2nd Battle Group 504th Parachute Infantry at Augsburg, which was redesignated the 1st Battle Group 19th Infantry, of the 24th Infantry Division (later the 24th Mech Div). My company commander was Captain Richard T. Pumphrey, a veteran of the 187th "Rakkasans" of the Korean War. He was an outstanding

---

2. See Bruce Palmer, Jr., *The 25-Year War: America's Military Role in Vietnam* (1984), for a biographical sketch of General Johnson, who survived the Bataan Death March. Chapter 2 of that book also includes material about Johnson and Palmer's time together on the Joint Chiefs of Staff during Vietnam.

soldier and man, dearly loved by all his officers and men. I served as executive officer of "Charlie 19," but the CO (commanding officer) was always off on special duty, so, as in Schweinfurt, I conducted the training, which meant I was essentially in command on a day-to-day basis and, inspired by Captain Pumphrey, I trained it to be one of the best in the division. We saw ourselves as some elite outfit. Captain Pumphrey gave me a lot of encouragement and a lot of leeway to do my job. I matured immensely as a soldier during that time, and few men in positions higher than mine showed as much concern for and leadership of me as Captain Pumphrey.

Our battle group commander was also a fine commander. I remember conducting training one day when one of the noncoms (non-commissioned officers) went "Ssst! Ssst!" and pointed over to some nearby bushes. I continued on but kept glancing over to the bushes, watching them shake and move. It turns out, the colonel was watching us on the QT from behind the bushes! We had a good laugh. When our colonel was promoted to chief of staff of the 24th Division, we got another good colonel, Colonel Schmerer.

One reason why the Pentomic organization was better than the triangular was the elimination of too many hierarchical layers of command. A much flatter organization was needed on this more modem battlefield. Businesses in the late twentieth century found out the same thing. The structure that replaced the Pentomic in the early sixties was call ROAD (Reorganization Objective Army Divisions), and it was big improvement over the Pentomic for one reason: it brought back the battalion. It also allowed for more flexibility.

Our battle group consisted of a bunch of paratroopers turned into mechanized infantry. At that time, I believe only the Israeli army had done something similar, and the result was an unbeatable elite force. In our first-alert position (that is, where we'd go if World War III started), we would become a combined arms task force with the mission of stopping an expected Soviet mechanized division. We had two hundred elite infantrymen, fifty combat engineers, numerous anti-tank weapons (including our first anti-tank missiles), twenty-five armored vehicles, several battalions of artillery support, tactical air support from the US Air Force, and more. We believed that if we could trade some space for time, our six- or seven-hundred-man force could inflict unacceptable losses on the twelve-thousand-man Soviet division that we believed would be attacking us. History has proven that these NATO forces of that period did provide a deterrent to a possible Third World War, and our successors continued to do so until the whole Communist system imploded upon itself.

We trained hard—so hard that later combat in Vietnam seemed easier in some ways. Our VII Corps commander, I believe it was General Clark, had some pretty firm ideas about tank-infantry tactics, and at one of the great German training areas, either Grafenwöhr or Hohenfels, he once put on a live-fire demonstration of a company tank-infantry task force in the attack. I was chosen to lead Charlie 19 as the demonstration unit, and we rehearsed until we got the timing down. On the chosen day, General Clark and a large entourage of guests gathered on a high hill to watch our blitzkrieg demo. We were psyched up and ready. Twenty tracked armored personnel carriers and five main battle tanks awaited my signal as the artillery, mortars, and air power pounded the objective. When the moment came, I gave the signal and roared out in my command track at the head of an impressive display of armor bounding across the terrain. Adrenaline was pumping as we roared up to a blacktop road with the tanks opening fire with their big guns. We all hit the road in a line at a good clip of speed and bounded over in the attack.

We thought things were going well as we pounded on down into the valley, but then the road suddenly broke up into a thousand pieces while the general and his guests watched. I knew timing was the whole thing, and so we roared up the objective, with the tanks peeking over the top and firing while the paratroopers bailed out of the APCs (armored personnel carriers) and raced up the objective with fixed bayonets, right there with the tanks where they should be. I'm told that the general up in the stands was beaming and exhorting his guests, "See, that's how it's supposed to work!"

I was directing the troops in the assault, when all of a sudden our left flank platoon started ricocheting a lot of rounds off a concrete pillbox into our midst. While I was sorting that out, a field grade officer came running up to me from behind, shouting, "You broke up the road! You broke up the road!" Our twenty-five armored vehicles had hit the road on line and probably broken up several hundred meters of it. That hadn't happened in rehearsals, and I didn't even know that it *could* happen. We were pretty excited, and the tank officers hadn't warned me. Anyway, we had done such a good job that I somehow got away with the road problem. I'm just glad I didn't have to pay for the repairs! No one has to convince me of the awesome shock action of a combined arms attack hitting an objective with perfect timing. Unless it can be broken up somehow en route, nothing can withstand it.

The training cycle restarted each year and included everything from retraining the skills of the individual soldier up to training larger and larger units, culminating in training on the big maneuvers. The result was that, if you were a motivated soldier, you just got better and better at your profession. I started out as a sharpshooter with the M1 rifle and as a marksman in some other infantry weapons in basic training, and by the time I was in the 19th four years later, I was an expert in all individual and crew-served weapons and the winner of the Expert Infantry Badge.

On one occasion we were conducting a training exercise at one of the great training areas on the subject "Rifle Company in Night Attack." It was a blank-fire exercise (no live rounds), and we dispatched one platoon to dig in on the objective and act as the aggressors. We had two or three young soldiers who had been Hungarian freedom fighters and were now earning US citizenship by serving in the army. These guys had a bit of an attitude problem because someone along the way had them thinking we were going to liberate Hungary, but that wasn't going to happen, especially with the odds something like three or four to one against us. While the platoon was digging in on this exercise, one of the Hungarians, Iha, found some live ammunition on the ground. A little later, a wild boar showed up rooting around at the base of the hill, and the Hungarian soldiers and a couple of others decided to have a barbeque. Iha fired a live round at the boar, wounding it, and then hid it in the brush, expecting it to stay there.

At suppertime, Sgt. Riviera, the platoon sergeant, marched up to the guilty parties. I was sitting on a water can with the company guidon spiked in the ground behind me, and I was furious.[3] We were about to have over two hundred men moving around in the dark for our night-attack exercise, but now we had a very dangerous wild boar loose out

---

3. Companies and battalions had guidons, which were small triangular flags of different colors (dark blue for infantry, red for artillery, etc.) and with the unit designation marked on them. They were mounted on a pole about ten feet tall. The guidon went where the unit went, usually attached to the leading troop carrier. It was carried in the front during parades and formations, and it sat in the company commander's office in the garrison. Like the "colors" of a battle group, the guidons were a source of pride.

there. I told the squad leader to take his culprits out and find the boar, bayonet it to death, and dig a six-by-six-by-six-foot hole and bury the boar. Out they went into the cold winter's night to find the animal, which they did, and Iha started for it with his bayonet fixed on his M1 rifle. The boar charged, the men panicked, and Iha swung his rifle and killed the boar with the butt, breaking the rifle in two. Sgt. Riviera radioed the story and we had to face the staffs on high about it.

Another memorable incident from that time happened when our battle group went on a maneuver that started with a long foot march in full combat gear. We marched something like forty miles and then went into a defensive position, falling back into successive defensive positions, a grueling exercise, and all on foot, with the vehicles being used only for weapons carriers and resupply, no troop transport. It was as realistic (as possible) an exercise of how it might be in battle, although we probably would have more mobility and mechanization. The maneuver scenario left us without decent meals along the way, and we were trying to catch bits of sleep when and where we could. It was winter, cold and raining, and it felt like an infantry soldier's hell. We were tired, hungry, and miserable—just like in war, except this wasn't war, so we were not also contending with fear.

Meanwhile, the infantry battle group had a heavy mortar battery of artillery [that was] organic to it, and they got to ride in their weapons carriers and, once in position, could find shelter or put up tents to get warm and dry. One of my buddies was in charge of a forward observer team in direct support of my company, and these guys rode in jeeps and made no bones about ribbing Charlie Company guys about the difference in our comfort levels. "Haw haw! You guys see this nice warm blanket? Want it? Hee hee hee."

One night about 2 a.m., we were struggling over a wide stream moving to our next position, and we were just about at the end of our endurance. Up came the artillery jeep with the artillery FO (forward officer) lieutenant on the passenger side, wrapped in a blanket and sound asleep. The jeep splashed into the water between two files of grubby, tired paratroopers, and it was mostly across the stream when suddenly a stray strand of concertina wire snapped up out of the water and caught on the lieutenant's blanket. With a mighty wrench it stripped him of his blanket and spun him around out of the jeep and into the drink. Charlie

Company let out a roar of righteous laughter as the officer thrashed around and then remounted his jeep. That livened us up for a while.

As we were getting into the next position later that night, hoping for some hot food and rest soon, I was summoned to receive a counter-attack order. When I later issued that order, standing with my officers at the edge of a forest in the dark of the early morning hours, I remember raising my binoculars to my eyes to study the ground in front, and then the next thing I knew I had fallen asleep on my feet and tipped face first, straight like a board, into a huge mud puddle. That time the laugh was on me.

I was proud of having earned my Expert Infantry Badge, and it turned out that our battle group had a large percentage of officers and men with the EIB. We were champions of the NATO shooting matches, and we stayed on our game, expecting the possibility of a Warsaw Pact invasion at any time. I would venture to say that during the half century of the Cold War, our battle group was probably as good a fighting outfit as the army ever had.

Our battle group was also the first US Army unit assigned the French SS.11, a type of wire-guided anti-tank missile. One of our best officers was assigned to run this outfit. He was a West Pointer and one who was a credit to his background. There was much controversy about this development, so in 1959 a competition was arranged in which our platoon would take on the best platoon of the 2nd Armored Division. They set up in front of a large stand of bleachers full of NATO dignitaries and many US officers and German panzer leaders. The tank platoon, made up of five M48 main battle tanks, deployed way down range and began a tactical advance toward the stands. At about two thousand meters, our guys fired inert missiles at the tanks and hit them all, even knocking the treads off one. The tankers then drove up in front of the bleachers and were subject to some hard questioning by the observers. Some of the senior armored officers simply refused to believe what they had seen and heard. It looked to me like if we had these SS.11 missiles in the weapons platoon of the rifle company, even a company-sized task force, such as we were, could wreak havoc against a much larger Soviet tank force, a move that would also disrupt their command and control. This weapon, if properly exploited, could preserve a lot of German real estate and stop the Soviets far short of their goal. The French SS.11 missiles may have been one of those things in the mix that helped keep the peace.

One day in 1959, our battle group was returning to Augsburg from one of the great training areas the army used in Germany, probably Grafenwöhr or Hohenfels. We traveled in full battle gear in a long tactical convoy, with vehicles some distance apart on the Autobahn. Air guards were required on each vehicle, since the war could start at any time and we could be caught on the highway. We were a pretty typical sight in Germany during those Cold War days. I was riding in the command jeep for Charlie Company up at the front, when suddenly we spotted a sleek silver jet fighter. It roared overhead, banked around way up the line of our vehicles, and then came screaming past the right side of our convoy at treetop level, probably for several kilometers. It caused quite a stir among us, as we saw the black German cross insignia on its side, the pilot waving as he passed. We Yanks had a strange feeling as we realized the Luftwaffe was back.

———————

When my three years in Germany ended, I was assigned to St. Louis. This was during the end of the Eisenhower administration, a time when the image of the "ugly American" prevailed abroad. Communism under Khrushchev was aggressive. Guerilla wars were popping up in many places of the world. Sensing the trend, I continued to prepare myself by searching out and studying this aspect of warfare. It was more of a forty-hour-per-week job than any I've had before or since.

I also did a lot of dating during this period and, in my mid-twenties, I had reached a place in my life when I felt a need to make some moral choices. I sat in my apartment one evening and decided to follow the Lord the best I could in the church of my fathers, the Catholic Church. I became what they called a "devout Catholic," and much of what followed I attribute to His guidance, including when I met my beautiful wife, Peggy.[4] I have always known it was Divine Providence.

While in St. Louis, I also had something of a political awakening. My family background on both sides was moderate Republican, and I stayed

———————

4. Peggy and I were married on July 15, 1961, and enjoyed our first days together in our cozy apartment in midtown St. Louis. I didn't realize it for a while, but being married and then, later, having my first child, changed my priorities. Peggy was also a great support to my career and developed an interest in world events to the extent that she was able to discuss such matters with the best of them in the officers' club.

that way until the late fifties, when I became disgruntled with the "ugly American" image that I felt misrepresented our country. In St. Louis, I found myself debating politics with some acquaintances, and by the time of the November 1960 presidential election, I had become a liberal Democrat and voted for John Kennedy. As the reader will see, however, I soon swung heavily to a position of strong disagreement with Kennedy and his advisors, not just because of what I observed during the early months of the administration but specifically because of their approach to Southeast Asia.

But that would come later. When Kennedy took office in 1961, I was like so many other Americans.[5] We were coming out of the golden fifties into a new era with a vigorous, youthful leader, and we were full of optimism for what lay ahead.

---

5. For an excellent discussion of the nation's mood at the time the Kennedy administration came to office in 1961, see Robert S. McNamara's *Argument Without End: In Search of Answers to the Vietnam Tragedy* (1999). Reading the description of the American mindset at that time renewed my memory of my own thoughts as a young officer during that time.

PART II

WAR DIARIES

CHAPTER TWO

INTO HISTORY—OVER WE GO

FALL 1961–MAY 1962

AS IT HAPPENED, in the fall of 1961, President Kennedy sent General Maxwell D. Taylor and advisor Walt Rostow on a fact-finding mission to Vietnam.[1] They came back with several recommendations, one of which was to stay out of the conflict there if the effort wasn't going to be a fight to the finish. The Kennedy administration decided to go into Vietnam with a strong advisory and logistical effort. The army started looking for eligible officers who could be assigned as advisors.

In late 1961, I, now a captain, received orders to report in early 1962 to the army's language school at the Presidio of Monterey, California. I would enroll in an eight-week course in Vietnamese before proceeding to assignment with the Vietnamese Airborne Brigade as a battalion advisor. That's what my wife and I thought the paper said. A premier assignment!

We were enthusiastic about the orders to Vietnam because we felt that our government had to do something meaningful to stop what

---

1. John M. Newman, *JFK and Vietnam: Deception, Intrigue, and the Struggle for Power* (1992), 125.

we perceived as the growing Communist threat. Peggy and I enjoyed a wonderful train ride to Los Angeles and then a scenic seaside train ride up the California coast to the Monterey area. We were fortunate to rent a unique studio-style apartment from a kind but eccentric older bachelor. The stay in Monterey was like a honeymoon for us. I discovered that I was one of a class of fifty-eight captains and first lieutenants and the only one with a wife along.

At the language school, I met up with a few of the Officer Candidate School classmates I had also served with in Germany, most notably Milt Craddock and Darryl Savage, two professional soldiers who, in my opinion, were among the best the army had to offer. Our instructor was a North Vietnamese refugee, and I learned a lot from him. At the end of the course I ended up being one of a few who could handle the language reasonably well. During my time at the Presidio and during my travel time to station, I also formed a friendship with a young West Point first lieutenant named Humbert Versace, whom I was told was from a well-to-do Eastern family. Versace was a serious young officer who liked to study his profession like I did, and once we were in the field he somehow took a lot of harassment from the West Point captains in our group. He and I were interested in spiritual things, and we went to church a lot together. He got an intelligence job that required paddling up and down a river in a small boat, and the story was that he was later caught by the enemy, tied to a tree in a village square, and executed.[2] I saw his name on The Wall in DC in May 1993, along with about fifteen of our guys.

As our language studies wrapped up, we awaited further orders, and we were sent out a few at a time from the Oakland Army Terminal or San Francisco International Airport with civilian passports. The real work was still to begin.

———

What follows are excerpts from letters I wrote to my wife and mother during this tour of duty. To make it easy for the reader, normal type is used for the

———

2. Editor's note: The official account of Versace's capture, in 1963, and execution, in 1965, can be found on Wikipedia, on the page titled "Humbert Roque Versace." He was posthumously awarded the Medal of Honor.

raw text that I wrote at the time, and italic type is used for editorial comments I've added while compiling this book.[3]

Oakland Army Terminal, 3 May 1962

We're still here. And they don't expect to be able to move us until 8–10 May.

This is really a snafu and has us all mad as hornets. They have some 70 officers alone piled up, with more dribbling in daily, just sitting around doing nothing. This is really bad when you stop and consider that orders were published in December [1961] and we received our port calls in February, stating that we would go on or around 30 April and that we would be in Saigon before 7 May.

This shows extremely poor planning, and the result cost-wise alone to the taxpayer will be a wasted loss of about $12,000 for one week's pay and allowances for 70 unproductive captains and majors. They have made no effort to put our time to good use around here, with the result that some guys are twiddling their thumbs in boredom, some have disappeared in San Francisco, some are drinking away the time, some sleeping it away, etc. A bunch of us found some good books in the library and we go work out in the gym daily and practice Vietnamese. We've got these guys who can't speak it snowed, and it's good for laughs. So far the inevitable card games have not cropped up and I'm staying clear when and if they do. Nobody seems to have any money to throw around, maybe that's why.

There's a big piece in the paper about Darryl Savage, OCS classmate of mine, who got over some time ago ahead of us. Savage was one of the two guys in our class who had trouble making captain because of his personality. The guy is ruthless about efficiency and apparently can't stand to work for somebody who doesn't measure up to his standards for soldiering. Needless to say, you have to be careful how you handle a situation like this with your boss in the army *(remember Support Center?)*, and Savage was too brash and so they burned him. But Savage is doing a bang-up job according to [journalist] Joseph Alsop *(no less!)*, who wrote the big two-column article about him. Savage is an advisor to the Self-Defense Corps in several villages of Vinh Binh [Vĩnh Bình]

---

3. Editor's note: Some minor changes to spelling, capitalization, and punctuation have also been made to aid in readability, and some editor's changes and notes appear in brackets.

Province.[4] He apparently has been working with his men long enough to produce a fighting force, and a successful one. In the past month his men have been hit by three major Viet Cong attacks, and he often goes out with them on combat patrols into Viet Cong zones, besides conducting normal training. Savage has been able to control Communist activity within his villages by isolating his people into the controlled villages you've been reading about. After a few vicious hand-to-hand fights using machetes, shotguns, and spears the Communists quit coming around in small groups.

A South Vietnamese army battalion under the command of a rough, tough Major Le Huan Thao is in the area to back up Savage's villagers, and there is close cooperation between the two. This last month the Viet Cong came back with three reinforced company-size attacks on Savage and his villagers. In the first, at Ma Lang, the village fell to the Reds, but Savage and his men were able to keep the Viet Cong from disengaging and escaping. In the meantime, Major Le Huan Thao's troops were able to come up, and they tore up those Viet Cong from head to toe. They had similar but less spectacular results in two other fights.

Savage and his buddies are doing a good job, so it appears. They have got the people fighting back and denying the Viet Cong support. This crucial factor has forced the Cong to launch bigger attacks to try to attain the same ends. And the simple fact is the more troops you throw into a fight, the harder it is to pull out of it and disappear. As it happened, the villagers under Savage apparently put out enough firepower to keep the Cong pinned into the fight once it was joined. *(The villagers have better weapons now.)* It's a time-consuming and costly business to try and break off a violent battle once it's started. And Savage and his boys held the Cong just long enough for the Vietnamese regulars to come up and plow into the Cong from an unexpected direction. The result? A massacre of the Communists. So the Cong is losing support *(there)* and is forced to fight costly actions to try to make up the difference. These two pointers usually spell death for guerillas, according to the book.

---

4. Editor's note: The spellings of Vietnamese names and words have been preserved as the author originally wrote them, sometimes with editorial notes following in brackets.

We are more convinced than ever that we are going to be doing what Savage is doing, [especially after talking to] some officers on their way back home coming through here.[5]

The story of a major we met at the officers' club on his way home is not so comforting, but it reflects the tribulations these guys had to encounter months ago. He was one of the unfortunates who pulled a full 18-month tour, arriving in Saigon in October of 1960, long before the buildup. He served his first six months on a staff job in Saigon, but by then the Red buildup was getting out of hand. Headquarters in Saigon grabbed every man they could spare and put them out in the jungle in an attempt to train the villagers to defend themselves. The major was flown by helicopter along with a large gathering of press people and dignitaries to a village that was regarded as very pro-government. After all the picture taking and handshaking, the helicopter flew off with the dignitaries, press, and security troops, leaving the major standing there alone with his kit bag, feeling *very* much alone. The major settled into his task of trying to build a fighting force from the village's Self-Defense Corps [later the South Vietnamese Popular Force] and found that he had a total of three shotguns *(!!!)* to work with. Every week, food and mail were airdropped or brought in by helicopter to the major, but on two occasions they forgot for a couple of weeks and he had to rough it. The appearance of that ration plane became a thing of great importance in the daily life of the major, for it was his only contact with America. One can imagine his feeling on the two occasions when it did not come. *(Oh yes! It happened to us later!)* If it was futile for the major to attempt to build a fighting force out of his villagers with their "heavy" firepower of three shotguns, he was at least able to set up a good warning system for alerting everyone when the Viet Cong was coming.

On three occasions the Cong did come swooping down on the village, and the major led his pathetic little crew out to set up ambushes. On each occasion, the villagers fired a few rounds from the shotguns and scattered into the jungle when the Cong opened up with burp guns, machine guns, and rifles. The major ran with the best of them, for there was

---

5. The following year, I encountered Savage as a classmate at the Advanced Officers Career Course at Ft. Benning. He had been wounded three times and had the Silver Star and a Combat Infantryman Badge. He was some soldier, but a handful of us wondered why the people in charge had denied combat decorations for others of us, telling us *we* were not in a war, even though we were there alongside peers who *did* receive such awards.

nothing else to do; once he tripped and cut his knee on a rock. He was a little disgusted with the whole situation as he told his story.

This is a very interesting situation. Americans have not operated anyplace like this since the colonial days with the French, British, and Indians. We will have to be tough and smart. We will have to speak their language and think and act like a native. We will have to be shrewd salesmen and top-notch soldiers. I will have no one but myself to rely on because we will have very few Americans nearby.

People *(back home)* don't seem to know a great deal about this situation because we don't want to build up a high-pressure war fever that will cause things to get out of hand. Some people don't like our being there because they don't understand and they don't like war. These things make it tough on us guys because we naturally want a little credit and patriotic support, like we saw our uncles get in WWII. Sometimes you get the feeling that we are bearing the brunt of this thing alone with our own families and that nobody else cares what really does happen to Vietnam in the end.

Oakland Army Terminal, 4 May 1962

The Vietnamese told us white men seem to be able to stand the jungle better than they (the natives!) can. This is good for us because if we can get into good shape we can gain advantages over the Cong. I read an article in May's *Popular Science* by a Special Forces sergeant in Vietnam. He says the Cong is afraid of the jungle and stays close to the roads and trails. He also says that although they set a lot of ambushes they do a poor job of setting them up and don't have the fighting power to stick around very long. This is more good news. If I determine that this is true when I get over there I am going to be able to devise ways and means to tear them up much easier than I thought. I thought we would have to take about 8 steps to hell to catch these clowns and shoot em up. Now it looks less hard.

This is quite a bunch of troopers I'm in with here. I can see already that we will leave one hell of a mark on Vietnam. Four of these clowns, Jim Lancaster, McGowan, and two others, got a taxicab driver into a crap game the other night and won his taxicab from him!!! It started out over the fare, I hear. They told the guy if he would chauffeur them everywhere they wanted to go as long as they were here, they would give him his cab back. They've been having this guy report at 0900 each morning and we all get free transportation anywhere.

A group of guys have been going to Tanforan Racetrack daily (right across from our motel, remember?), and they've been cleaning the place out. Remember the clown who took Las Vegas for $1,700? What a bunch! They are going to be sorry they didn't ship us out right away.

John Rodgers got himself into a fix a couple of days ago, though. He locked himself out of his room. After trying every available means to get back in, the guy in the room next to him, Norm Garner, came along. You could hear the bellows of anger down the hall. It happened that there was a door between their two rooms, so Norm let John in. Then Norm locked up his room and took off. John promptly proceeded to bolt the door to Norm's room and with a cry of triumph he stepped out of his own room and promptly locked himself out again. We were all watching the situation by now. John walked about 10 steps up the hall, grinning like a cat, when he reached into his pocket and no key!! A look of horror crossed his face as he jerked to a stop and pulled out his pockets. NO KEY! He dived back to the door and shook the doorknob frantically. It wouldn't open. Such nasty language, tsk tsk. Several hours later he was able to get someone to let him in. I think he wears the key on his dog tags now. *(Handpicked we were!?!)*

I went with another captain on a Pan Am civilian jet. What they were trying to prove was beyond me—they sure were not fooling anybody. Not a good omen. We made a 24-hour stop-over in Honolulu, where we were put up in a hotel on the beach. My partner and I were in uniform and were solicited by the people who were filming the TV detective story *Hawiian Eye.* [They asked us] to walk in front of the camera for a sequence they were shooting. Big stars! Little did we realize that the little interlude in Honolulu would be the last decent life we would know for months.

I'd like now to say something about who was signing up to go to Vietnam in 1961 and how popular opinion on the matter differed from what I saw on the ground.

First, Colonel David Hackworth, the most decorated soldier from the war and, at this writing [in the 1990s] a contributing editor to Newsweek, *was quoted in 1992 by historian Howard Means as follows:*

"When they asked for volunteers in 1961, to me the writing was on the wall. Go to Vietnam. It was very important to get that experience. Those young first lieutenants and captains who went to Vietnam in 1962 were all walk-on-water types—they only picked water-walkers. They wanted to give them battle experience so that later, when they were three- and four-star generals, they at least would have some combat experience. . . . We sent them to Vietnam so they would be ready for the real war we were going to fight with the Soviets." [6]

Shelby Stanton, a military historian, was also quoted by Means:

"The idea was to get some kind of backbone into the above-quality units of the Vietnamese military so that there would be something to look up to, something that would enhance profession-alism in the military." To have been selected for the job, Stanton says, "means that you're considered among your peers above aver-age in endurance and toughness, but you have to have more than a fundamental knowledge of patrolling, infantry tactics, and how to set up weapons. You also have to be able to work with people . . . . It takes a very political person. You need the utmost consideration in knowing when to apply pressure because it's another culture . . . . But when the military has an excess of good officers as they did in the early days of Vietnam, before they had to pull a lot of American troops over there,[7] you would find they really did sort of tag people as far as assignments went. The fact that certain people would be slotted for certain assignments went to the fact that they had not only outstanding attributes but personalities that befit the job..." [8]

William Prochnau in his book Once Upon a Distant War *(1996) shared a similar sentiment:*

---

6. David Hackworth quoted in Howard Means, *Colin Powell: Soldier/Statesman, Statesman/Soldier* (1992).

7. The press reported, I believe in 1965, that in the previous eighteen months the army had lost twenty-eight thousand officers with between three and thirteen years of experience. This had to have had a serious impact on the troops then committed to battle.

8. Shelby Stanton quoted in Means, *Colin Powell.*

Rarely was an American army ready, and the early carnage in the country's wars was often terrible. But this force was different. It was professional, elite, ambitious, and self-certain, the advance guard of a new empire born out of The Bomb and the wreckage of two world wars in thirty years. Of the 11,200 American military men in Vietnam by the end of the year [1962], not one was a draftee, few if any doubted their right or need to be there, and most saw their tour not as dirty duty but as a desirable career opportunity. Vietnam became a handy tropical practice field— "the greatest continuing war games we have come up with," an officer told [war correspondent David] Halberstam.

The Vietnam of 1962 drew the best the United States had to offer—sons of its most famed Second World War generals, top-echelon graduates of West Point, bright men from nowhere who had risen from one of America's greatest mid-century strengths, its openness to merit. Its warriors had attended the best staff schools, studied for advanced degrees in philosophy as well as military science. These were men on the rise who intended to get ahead in their profession, and Vietnam was the place to be. "The word is ours," the officer continued. "If you want to stay on the ball team, you go to Vietnam."

In the roughhouse of career advancement, the military takes a backseat to no institution—not to the brutal dueling of corporate life, not to the deadly character assassination of politics, not to the ruthless candlelit conniving of the church. So the corps of young officers on the move included all the usual players—the innovators, the strategists, the heroes and the do-gooders, the cynics and the hustlers, the sycophants, the men who would dig their boots into the ribs of their best friend to reach a higher handhold on the climb. Out of this group would come the military's next generation of leadership.[9]

These are flattering analyses from Hackworth, Stanton, and Prochnau, but I really think the military initially looked not for top talent but for stateside infantry first lieutenants and captains who were just about

---

9. William Prochnau, *Once Upon a Distant War: David Halberstam, Neil Sheehan, Peter Arnett—Young War Correspondents and Their Early Vietnam Battles* (1996).

eligible for an Asian tour. I thought I was specially qualified, but I don't think the army cared about that, or even knew it.

In mid-May of 1962, it was finally my turn. When we arrived at Tan Son Nhut Air Base near Saigon, we were taken to an area called Tent City. This was the in-processing area for our American buildup. At the time we arrived, there were still only a few thousand Americans in Vietnam. I discovered that in our top-heavy American bureaucratic way, the army had already superimposed Headquarters MACV [Military Assistance Command, Vietnam] on top of MAAG Vietnam [Military Assistance Advisory Group, Vietnam]. Although MACV was a joint-service head-quarters with a specific battle zone, the Pentagon did not declare MACV a full-fledged theater command and placed its operational jurisdiction under very tight restrictions. COMUSMACV [United States Military Assistance Command, Vietnam] reported to the Commander-inChief, Pacific (CINCPAC), who in turn received his orders from the Joint Chiefs of Staff.

MACV operated under strict geographical limitations as well. For example, special permission from Washington was required for operations involving North Vietnam or Laos.[10] This was an obvious recipe for defeat, and top-grade soldiers worth their salt would scream to high heaven before they would let us get put into a war that way. Everywhere in Saigon were field grade officers in starched khakis, new building going on, and of course air conditioners going into the "key" spots. Even many years later I can remember that my first impression of seeing this really dismayed me; I knew that something was seriously wrong. There was, of course, a spiffy officers' club on the top of a tall new building in downtown Saigon, and I would later visit it from time to time, when I came in from the field. Sometimes in the evenings we could watch fighting going on off in the distance as we enjoyed an after-dinner drink. I would think, "Enjoy it while you can. You'll be back at it yourself in a day or two."

A lot of politicking was going on as new arrivals tried to line up the best possible jobs for their careers. The field grades in Saigon were mainly West Pointers, and they were looking after the West Point officers who arrived with our group. Those of us who were not West Pointers called it the WPPA, or West Point Protective Association. Incidentally, in the midst of

---

10. *Webster's New World Dictionary of the Vietnam War* (1999), 257–58.

all this I met General Patton's son, George, who was a field grade officer on his first tour in Vietnam. At the time he looked more like a bureaucratic type than the fighting image his father had left behind, but I understand he went back later and had a fighting assignment.

One of the things I noticed during this period in the army was the large number of officers, especially higher-ranking officers. Headquarters seemed more and more top heavy the higher up one went. Enormous amounts of busywork were being generated and dumped on the relatively few line units, and staff officers couldn't seem to leave their hands off of lower commands and were micromanaging everything. This problem carried over to the Vietnam War to the extent that even senior officers became concerned about it. General Bruce Palmer, Jr., for instance, wrote of his days at the Corps Command level, "The routine use of the helicopter as an aerial command and reconnaissance vehicle by many senior American tactical commanders resulted at times in gross oversupervision of, and sometimes unwarranted interference with, subordinate commanders, in particular company and battalion commanders. The results were not very good. Overzealous leadership at division level also resulted in bypassing intermediate commanders, further weakening the chain of command."[11] This didn't happen in the advisory effort, as far as I was concerned, but it was a huge problem in my division in Germany.

Company officers in line units were regularly pulled out onto all sorts of "SD" or special duty assignments, leaving behind one officer to run the company. In my situation in Germany, that one officer was me. So I trained the units and led them in most of their training operations, and I think my superiors thought I did a good job, so it helped my career immensely. It made a good fighting leader out of me, and I turned over a well-trained outfit to my superiors and successors. But it wasn't good for the army as a whole. I noticed when we were performing night-combat training in the cold winter in Germany that the micromanagers were not around. Neither were they around deep in the mountain highlands during combat in Vietnam.

Historians Gabriel and Metz wrote about top-heavy militaries in From Sumer to Rome: The Military Capabilities of Ancient Armies *(1991):*

---

11. Bruce Palmer, Jr., *The 25-Year War: America's Military Role in Vietnam* (1984).

> *The relative strength of an army's fighting forces to its adminis-
> trative and logistical tail is called the "teeth-to-tail" ratio . . . . As
> every soldier knows, large logistical tails and administrative staffs
> breed officers who must be attended by still other soldiers. Indeed,
> there seems to be a rough historical rule that the effectiveness rate
> of armies is inversely proportional to the number of officers relative
> to troop strength. The most combat-effective armies in history have
> had the smallest officer corps. Thus the Roman officer corps rarely
> exceeded 3 percent in strength; the Prussian army of Frederick, less
> than 4 percent; and the German army in both world wars, under
> 6 percent. By contrast, modern armies are top-heavy with officers.
> The American and Russian armies [circa 1991] number approx-
> imately 12 and 14 percent officers, respectively. Only the Israeli
> Defense Force, commonly regarded as a very effective battlefield
> army, with 6.5 percent officers, approaches the officer strengths of
> ancient armies.[12]*

*I caught on to this concept very early and profited from staying with
the troops as an operations and training officer. This is why, on arrival in
Vietnam, and seeing the fat Headquarters MACV superimposed on top of
MAAG Vietnam, I was dismayed and sensed trouble. I had rationalized
the top-heavy nature of the army of the fifties and early sixties prior to this
as a kind of response to the threat of nuclear war. If we lost large numbers of
ranking officers in the initial exchange of nuclear weapons, we had ready-
made replacement cadres populating all these top-heavy headquarters. But
I knew deep inside that we were going to have to be lean and mean and in
fighting trim to make it in Southeast Asia, and here we were starting out fat
and top-heavy. I lost a lot of respect for more senior people as I watched this.
This was an important factor in what ended up happening in Vietnam, in
my opinion.*

Tan Son Nhut, Saigon, Vietnam, 14 May 1962
It was mighty hot when we got off the plane, reminded me of Ft. Benning. They moved us by bus to Tan Son Nhut Air Force Base (Vietnamese), where we are bedded down for the present in a fortified tent camp.
We are located directly west of Saigon about 30 minutes from town.

---

12. Richard A. Gabriel and Karen S. Metz, *From Sumer to Rome: The Military Capabilities
of Ancient Armies* (1991), 88–89.

The going is rough as we expected. The normal things of life come hard here for us. It's difficult to get enough water for any purpose around here. And it's so hot you need plenty. We have to walk around a day and a half without wash water and not enough to drink to be satisfied. This is because the facilities here are so overtaxed. Too many of us hitting this place all at once. But it's necessary. They have guards all around this place with sandbag fortifications and barbed wire. Right next to us is a helicopter outfit that moves out against the Cong daily. They got clobbered yesterday. I got some films of them taking off, which you will be getting one of these days. Those very choppers that you will see in the film I'm sending you got shot up pretty bad by the Cong on that mission. I'll be moving out of here in a few days to my new job. *(I was WPPA'd out of my assignment to the airborne brigade by an older West Point captain who was a classmate of mine at the Presidio. He may have lived to regret it, as I will recount later.[13])*

. . . Rumor has it I'm going to the II Corps to be a battalion advisor. Saigon is OK; the people drive like nuts and they get excited when we talk Vietnamese to them. We find we can speak Vietnamese a lot better than we thought, and we attract crowds when we do. White men never bothered much to learn their language, which is one reason why we are fighting here now. The general *(I think Timmes[14])* introduced himself to us and gave us a pep talk today. They tell us they picked us by name for this job. I wonder who my buddy was! I ran into some higher-ranking people I knew in Germany and was offered my choice of jobs Sunday. *(Ha, ha! All that went up in smoke!)*

Tan Son Nhut, Saigon, Vietnam, 15 May 1962

The *(H-34)* helicopters next door are revving up their engines preparing to go out on a mission. We are waiting for transportation to go into Saigon to finish our processing here. Some of those choppers are really patched up. *(This particular chopper outfit appears to have been an early CIA outfit brought in possibly as a part of Air America.)*

---

13. WPPA (West Point Protective Association) was a slang term used by officers who had acquired their commissions by means *other than* West Point, and it expressed our dissatisfaction over repeated alleged efforts of West Point graduates in positions of influence to unfairly help the careers of other West Point graduates.

14. Major General Charles J. Timmes, MAAG chief from July 1962 to May 1964.

The Cong has gone to work on the railroads for some time. But last week they blew up a train and chopped up the tracks around here pretty bad. One of my classmates at Monterey, Anderson, is being assigned in charge of railroad security. That's a very interesting and frustrating job. He has to keep thinking up ways to keep the Cong from blowing up trains, and the Cong keep trying to outthink him. It becomes quite a game between the two, with human life the stakes. He will probably use methods developed by the Germans in World War II. *(Well, not all of them.)*

A word about helicopters: L. Fletcher Prouty in his 1992 book JFK: The CIA, Vietnam and the Plot to Assassinate John F. Kennedy *was knocking helicopters as expensive battle taxis (him being an Air Force jet pilot type), and he also knocked the military industrial complex for its role in selling all this stuff to the government. Prouty's theme in the book is pretty shocking, but what I wanted to point out here is this quote from the chapter "The Battlefield and the Tactics": "Helicopter losses were staggering. 'Of the 6,414 total aircraft-related deaths, to April 17, 1971, . . . 4,622 [occurred] in [helicopters],' according to a U.S. Air Force policy letter from May 1971."*

An even more shocking statistic from the same policy letter follows: "Of the 4,622 deaths in helicopter crashes, 1,981, or 43 percent, were 'casualties not from actions by hostile forces.'" Prouty says, "If you had helicopters, you did not need an enemy."

He goes on to say that between 1960 and 1962, "when the American military advisory strength in Vietnam was limited to 16,000 men, Gen. Paul Harkins, then the senior commander in Saigon, complained bitterly that with a ceiling of 16,000 men, he could get only 1,200–1,600 effective combat advisors, because most of the rest were confined to logistical support work. The bulk of that support work was related to the helicopter."

That may have been true, but the helicopter company assigned to our area out of Pleiku not only flew many of the resupply missions to the advisors that I relate later, but they were also used tactically to move Vietnamese troops into combat operations, which I also relate later. I served in both parts of the thing: as part of the logistical tail, still dangerously exposed to combat, and later as the combat advisor.

Tan Son Nhut, Saigon, 16 May 1962

Got my assignment yesterday. I am to command all American elements at Kontum [Kon Tum]. Kontum is located north of Ban Me Thuot

[Buôn Ma Thuột] and Pleiku on a key road net *(Highway 14)*. The terrain is mountainous jungle. It is near the Laotian and Cambodian borders in the central part of Vietnam. Henry and Williams are going with me. The Vietnamese 40th Infantry Regiment and several divisional units *(of the Vietnam army's 22nd Division)* are there. Kontum is located close to the Ho Chi Minh Trail. Although it has been quiet there for the last few months, it was frequently attacked by the Viet Minh when the French were there and up to a few months ago. Then the Vietnamese army moved in the 40th Regiment, and that's too much for the Cong right now. The Cong keep smacking at Kontum whenever they can because they are afraid the troops at Kontum might attack the Ho Chi Minh Trail and cut off their route of supplies. Yet since the 40th moved into Kontum I don't think they have attempted to attack the trail. I don't know why, but I guess I'm going to find out. Maybe it's because they don't want to move into Laos, but in view of circumstances in Laos as of now, it would seem to me Vietnamese troops would be welcome. *(Note from 1997: I was thinking in terms of a uniform SEATO strategy for the region, and I assumed at the time our leaders would certainly be thinking that way.)* In any case, I guess the Cong can't muster enough strength to hit Kontum, Pleiku, Ban Me Thuot and other VN strongholds either, so action is limited to minor skirmishes. That's bad enough when it's you that gets involved. *(This situation changed only weeks later, when the government gave away Laos, a huge strategic error for the US. Then the whole flank was open and the enemy could and did come over in ever-increasing numbers.)*

For a good story about the Kontum area I recommend reading *Street Without Joy* (1961) by Bernard B. Fall, who gives an account of the French fighting there. Some of the current fighting around there *(circa 1961–62)* has been similar, and the reasons for fighting in that area remain the same. Descriptions of what the country is like are good.

The monsoon or rainy season has not arrived yet. It's due any day. The people here in the firmly held Saigon area are mostly friendly. Their attempts at English make us smile as much as our attempts at Vietnamese make them smile. But we are finding that we are actually pretty good at the language. They really get excited when we start speaking Viet to them. Americans who can't speak it seem like islands lost in the great Pacific; nobody pays attention to them [the Vietnamese] or just tries to communicate with them if spoken to. Too many of these

(American) clods treat the people rough, and that's bad. But when a white man speaks the language—and we are among the first—wow! We gather crowds, smiles, and many friends everywhere. They vie with each other to do favors for us, and they go out of their way to come over and speak with us. *(How things changed a few years later.)* I notice the peasant types are very withdrawn, sullen, and frigid toward all white men. The Cong put them that way. *(In later years, we put them that way.)* But most of these people even thaw out with a big grin when we stop and politely speak to them. The value of a few Americans like us, we discovered, sent over here to speak the language, is fantastic to American hopes here. We would be worth our weight in gold if we just walked around the streets shooting the breeze with these people. *(Do you suppose that was pretty naive?)*

I do not like our tactical security here in the Saigon area. We rely almost entirely upon the Vietnamese police and army for protection and our fortification; defenses and security measures are inadequate. Many of our people could get shot up until they could come to our assistance. We've done a few things, but not enough. Especially in view of the fact that the Cong made an announcement that they would like to nail a few Americans for Ho Chi Minh's birthday on 19 May. We have barbed wire and firing emplacements sandbagged around here. But the layout of the stuff is not good, and people have not been told of any defense plan. Sentries are everywhere, but they don't seem to be on their toes. The buses have wire screens over the windows to prevent grenades being thrown in—that's good. But security of trucks used to haul people is poor. Yesterday 20 of us sat in the back of a truck in downtown Saigon, stationary for 30 minutes. Most of the guys were batting the breeze like they were back home. Not some of us. Rodgers, Henry, and I got the hell out of that thing until it was ready to move. It stayed sitting around one place plenty long enough for someone to summon a Cong agent to toss in a grenade. Open vehicles must *not* sit around full of personnel in a place like this for more than a few minutes.

Another thing, these Americans around here are repeating the mistake of the French by talking too much about military information in places where the wrong ears can hear. This is probably one good reason why the Cong evaporates when we launch an attack on them. We Americans are not ready enough to fight if we should have to more than we are. A sudden shock attack by the Cong could wipe out a lot of us who could be saved if proper security measures were used. I smell trouble in the air. This morning, even though it is none of my official business, I grabbed one of the top sergeants and squared him away on these guards around here. Every damn guard should man his post like the Cong is after him. *(I saw this type of problem cost lives later, up in the Central Highlands.)* Yesterday John Rodgers and I told a bunch of blabbermouths to shut up and why. I'm a hardnose here because that's the way it has to be in a combat zone. I'm going to get out of here, and my people are going to get out of here, you can rest assured. I don't mind banging a couple of heads together to see to it.

My personality undergoes a Dr. Jekyll and Mr. Hyde change when I get around troops. They don't get ripped. They realize it's for their own good in the end. "Good Joes" get guys hurt when the chips are down in this business.

Saigon has many types of buildings, from the modern type to little thatched-roof huts. Streets are very similar to any city, but the traffic consists of every imaginable type of conveyance. They have Cadillacs and every kind of automobile. They have bicycle taxis, horsedrawn carts, three-wheeled bicycle taxis, motor scooter and cycle types which defy description. They drive everything like maniacs. You get a thrill a second in a taxicab around here.

The wealthier district has some amazing homes of architecture I've never seen before. It must be a cross between European and Vietnamese architecture. These type homes go in brilliant colors. The government buildings are European style, a yellowish stucco with red tile roofs. President Diem's Freedom Palace [Independence Palace] still has the right wing bombed out from that sneak attack by those two pilots, and I was surprised to see how wrecked it looks.

There are shops and stores of all descriptions, from the most modern American type down to sidewalk vendors. You can really haggle prices with these people, and I like to go with a bunch of the boys just to argue prices. You express interest in something and forty salesmen descend upon you. You ask how much and they tell you and go into gyrations because you speak the lingo. People start to gather around at this strange phenomenon of Americans who speak Vietnamese. You tell the forty salesmen some outrageously cheap price you'll give him and the argument is on. We finally give up on the guy and move on. The forty salesmen rush up and down the street spreading the news that there are Americans around who speak VN. Often these bargaining sessions become a highly animated question-and-answer session, and everyone except the head salesman forgets what was being sold. It must be quite a sight, two or three Americans, even me towering over a crowd of Vietnamese, and all of us jabbering up a storm in VN.

There is a slight tinge of fear and tension that grips the air here. You stay on your toes and sleep on your loaded weapon. You sleep under a well-tucked-in mosquito net, and even the little devils sneak in. You wash as often as possible, and that's not enough. You watch like a hawk where you get your water from. You shake your boots out and look em over before you climb into them; the same thing with anything you touch. You watch the local citizens that gather around you with a suspicious eyeball. You check over the vehicle you get in. You search the area with a careful gaze before you walk out of a building into the street. You are guarded about what you say and where you say it. It ain't like home.

I'm a relatively small American, but I'm a brute compared to these people. They are small, tough, wiry, and often undernourished. Many of them are stuck with one or more diseases. Their needs seem impossible. Education is needed. The food supply is generally adequate, but they don't eat the right stuff. Sanitation needs are basic and simple, but how do you change an ingrained culture of poor sanitation habits that are a centuries-old ritual?

Here is some general context about Vietnam's geography, weather, wildlife, and people: Vietnam is 127,000 square miles, about 80 percent the size of California, or twice the size of Missouri. It has about a 1,500-mile coastline, whereas California's is 1,000 miles. In width, the country varies from 35 to 300 miles. South Vietnam itself is just a little smaller in total area than Missouri.

This is a country where the terrain actually dominates the people rather than the other way around, as in the United States. The most dominating feature of the area is the Annamite cordillera, which spreads over most of the country. This chain of mountain ranges also affects the entire border with China and Laos, and much of Cambodia. Virtually all of this vast area is covered with jungle. The mountains range in elevation from 2,000 to 10,300 feet above sea level. I think perhaps the average runs between 2,500 and 4,500 feet.

The second major geographical feature of South Vietnam is the Mekong Plain and Delta. It is in the extreme south of the country and also extends to much of the adjacent border territory with Cambodia. Named after the Mekong River, one of the world's great rivers, this region is low, swampy jungle and at times nearly impenetrable in regions not inhabited.

The third and last major geographical subdivision of Vietnam is the coastal plain that runs along the South China Sea and the Gulf of Tonkin in the north. To the west, this plain meets the Annamite cordillera. At many places, the mountains extend right into the sea, but at most places there are at least a few miles of width to the coastal plain. The widest points get to the neighborhood of about 70 miles. Most of this terrain is not dominated by the jungle, or at least that which I have seen is not.

I think I could sum up the weather situation in Vietnam by mentioning that there are basically two seasons: wet, or monsoon, and dry. The exact nature of how these seasons appear seems to depend on where in the country you are and how far above sea level you are. The monsoon or rainy season comes at a completely different time in the mountains than in the Mekong Delta. In the highlands, the dry season can actually produce some very chilly weather in the 50s for a month or so in January. (That was an unpleasant surprise for us, as we had hot-weather gear and had gotten acclimated to the hot weather.) In this region, the highlands, the dry season runs from about November to May. There is virtually no rain at all. Things become extremely parched and dry, and water levels reach near drought level. Dirt roads and trails become covered with several inches of dust, and you must travel with a bandana over your nose and mouth. You are literally coated with dust in this season. (Tough on weapons, aircraft, and vehicles!) All forms of life seek water and often come into collision with man. (Great surprise for us Yanks.) We have actually been engulfed in clouds of billions of flies coming out of the jungle looking for water. You can see them coming as dark clouds moving across the terrain, like a biblical plague.

The wet season produces frequent rainfall that grows more intense toward late May and lasts longer each day, then tapering off until the monsoons end in November. This is the highlands I am speaking of. In this season, everything turns into a quagmire. The same dirt roads that had us nearly axle deep in dust are now an unbelievable, almost impossible mass of muddy clay. You have to have special drainage ditches to carry off the water, as we learned the hard way. The rain comes down so hard at times that several inches of water can be seen rushing across the ground. You almost feel that you are sinking at times. When it gets really wet for some time, many forms of life, especially snakes, seek temporary dry spots and, once again, they collide with man. Streams and even rather large rivers that actually dry up during the dry season become raging currents. Flash floods occur when perhaps a violent thunderstorm miles away somewhere in the mountains will send a deluge of water cascading downward on unsuspecting areas. We lost ninety men out of a battalion one day, caught in such a flood. When the water receded, the Viet Cong our men were pursuing returned to collect whatever weapons and equipment they could, and we lost plenty. Some of the survivors were in the water so long the entire skin of their bodies looked like "dishpan hands." This is a glimpse of the climate in Vietnam. I mention these things so that you can gain a small appreciation of the problems we faced.

Now what about the animal life in Vietnam? There are in the highlands some elephants, many deer, tigers, panthers, and various breeds of wildcats. Most of this type of wildlife avoided us. Unfortunately, the snakes and insects did not avoid us so much, probably because they were not smart enough to. Whenever you ran into one of these monsters, both of you did your best to beat it out of there. But unfortunately, snakes, tarantulas, and scorpions blundered into your quarters now and again, and then you had a battle to the death. Snake varieties included the python, king cobra, and numerous varieties of highly poisonous pit vipers. There were also endless populations of large rats, everywhere so common I almost forgot to mention them. In the jungle itself, another common problem were the large blood-sucking leeches that attached themselves to everyone. It all probably sounds horrible, but you learn to live with it immediately, even as we do our traffic hazards in the States.

And now a word about the people of Vietnam. The total population is about 30 million. About 14 million persons lived in South Vietnam circa 1963. Of those, probably a little over a million were Montagnards (French for "mountain men"). [Editor's note: Today they are more often known as a group as đồng bào Tây Nguyên, *which means Central Highlands compatriots, or* dân tộc thiểu số Tây Nguyên, *Central Highlands minorities.] They live in mountainous regions as tribes. These people closely resemble Native Americans and have received much the same kind of treatment from the Vietnamese that Americans have given our Natives. The Montagnards are quite primitive and simple, but they are a fine bunch of people. The Vietnamese dominate the Delta, coast, and highland areas.*

CHAPTER THREE

THE EARLY DAYS

MAY–JULY 1962

I WAS FLYING over the Mekong Delta, in the southern part of Vietnam. The countryside was flat, under a heavy canopy of foliage from the rice paddies. There were roads and numerous towns and settlements. As I looked down, aware of the kind of war going on, I did not realize then, as I did later, how this scene would eventually become so familiar and so tragic for so many of my countrymen. As the plane moved on into the mountain highlands, the terrain became rugged, covered by forest, and with fewer roads and settlements. After a few stops, I believe at Ban Me Thuot [Buôn Ma Thuột] and Pleiku, the C-123 arrived at Kontum [Kon Tum]. Pleiku was the only landing field with the ability to home in a plane electronically, so they couldn't land anywhere else in bad weather, a fact that affected our lives in that area for the duration.

Kontum, a provincial capital, was like a town from a Western movie. The base camp, located at the northern edge of town, was not what I expected at all. It was like a kind of a fortified resort motel. Literally days before I arrived only a few Americans were there, but now, suddenly, scores of us were arriving to beef up the advisory and support effort. Headquarters of the 221nd Division of the ARVN (the Army of the Republic of Vietnam) was in the town, as were a number of combat outfits from that division. At the time, a full bird colonel commanded a Viet division, and a US lieutenant colonel, Anthony J. Tencza, Polish American like me, was his senior advisor. Tencza lived in the base camp and was the ranking American there. Kontum Province was roughly the size of New Jersey, and the province chief (the equivalent of a US state governor) was a captain. His top soldier, who commanded a brigade-sized force comprised of Vietnamese Civil Guard and Self-Defense Corps troops, was also a captain. They later made me his senior advisor. That one got past the West Point Protective Association and was probably as fine an assignment as a young captain could hope for, the normal equivalent for a brigadier.[1] I was the base camp commander, but Col. Tencza was the senior officer resident, so I worked for him. He had jumped with the 82nd Airborne in the D-Day Normandy night jump and also with the 187th Rakkasans in Korea—a real fighting type.

The day I arrived, my job was to relieve the former base commander, an older, tall, thin black captain with a French name who was due to go home. In some ways, the job was similar to that of headquarters commandant in a US outfit. Advisory teams, at that time consisting of a captain and one or two NCOs (non-commissioned officers), were quartered at the base camp. There were about 150 officers and NCOs, and their job was to advise the Vietnamese battalions. When the Viet units went into combat operations, the advisor teams were gone with them, leaving only a handful of support troops to run the base camp and support the others. To protect the base camp, I had under my command a Vietnamese sergeant and a platoon of Montagnard soldiers from the Civil Guard, but at any given time the place was nearly empty, as most people were out on operations. We had a small, single-side band radio detachment that maintained contact with the outside world.

In our support duties, we often took two choppers or two vehicles out to resupply our guys or retrieve casualties, and as it ended up, the support job was often more dangerous than combat operations. At that point in the war, the enemy would usually avoid a fight with an infantry battalion, but they sought to ambush small resupply missions such as ours. It was our perception that I personally should accompany as many of these missions as possible, to make a statement to the advisory teams in the battalions that the base camp cared.

---

1. That ended up being the equivalent of a brigade job, a very good job for a young captain, and being there probably got me the job before someone else noticed it.

We also had to contend with the helicopters, which was its own liability. Pleiku had a chopper company of obsolete YH-21 "flying bananas," which were woefully underpowered for the mountains. The heroes in that outfit seemed never to get sleep, as they were always maintaining the aircraft or flying, often in harm's way.

The logistical rules for stocking the camp were peculiar at that time. We had to plan out mess requirements and buy the supplies (food, etc.) from the navy in Saigon. As individuals, we were paid a special type of allowance, and to cover mess hall expenses, we collected dues from everyone, officers included, forming what the army called a non-appropriated fund. It turned out to be something of a small business operation. As the base camp commander, I found myself in the role of business manager and accountant of the operation. Instead of using the army's G-4 logistics setup to make sure everyone had what they needed—a setup that was maintained at great cost for just such a moment, and that we were all trained to use—someone in charge had instead set up this quixotic system of non-appropriated funds, which just added burden to the troops and was too fragile for a combat situation. There were a lot of things wrong with this, but I especially remember that about the time I came in to replace my predecessor, a plane load of supplies flown up by the Air Force and worth over $4,000 disappeared. They made the lot of us pay for it.

Kontum, Vietnam, 18 May 1962

After a bouncy flight in an Air Force C-123 with one bad engine, we were mighty glad to hit Kontum.

Kontum is the capital of the province by the same name. The province is about the size of New Jersey, and the capital, home to about twenty thousand people or so at the time, looked something like an old Hollywood Western town. Kontum is part of an area called the Central Highlands, which is an isolated plateau at the southern end of the Truong Son [Trường Sơn, or Annamite] range in west central Vietnam, bordering on Laos and Cambodia. It is about twenty thousand square miles of mountainous wilderness covered with tropical and bamboo forest and occasional tea, tobacco, and coffee plantations. The principal towns at the time were Ban Me Thuot, Pleiku, Kontum, and Dak To [Đắk Tô], all scenes of military activity and fighting at different times throughout the long war. The population was mainly Montagnards (highlanders) in the hills and countryside, with Vietnamese in the towns. There were about eight hundred thousand Montagnards in the area when the war started. An estimated two hundred thousand died in the war fighting for their homes, and most of the six hundred thousand survivors became refugees.[2] Although it would appear there has been little American concern about them, those of us on the ground know that the Montagnards, our staunch supporters and possibly our toughest fighting comrades, suffered a critical blow as a people in the war. In the II Corps area (which had headquarters at Pleiku, forty miles south down Route 14), Kontum was headquarters for the 22nd Division of the ARVN (Army of the Republic of Vietnam). Ten years later, in 1972, when the North Vietnamese Army (NVA) launched their Eastertide Offensive (with US troops gone from the area), Kontum was the site of a major battle in which the ARVN 22nd and 23rd divisions successfully beat back the NVA. Kontum was nearly destroyed by the heavy fighting and B-52 bombing. Finally, having been abandoned by the Americans, in March 1975 Kontum was evacuated by the ARVN at the beginning of the final defeat. The North Vietnamese saw the Central Highlands as key to the conquest of the south, and using the Ho Chi Minh Trail to establish bases in Laos and Cambodia, they planned to attack eastward to the sea, cutting the south in two. II Corps was responsible for the defense of the central portion of South Vietnam.

The American group here is fairly large, having grown like an explosion over the last few months. We have a fortified compound outside on the north side of Kontum all to ourselves, and our people move out daily to work with the Vietnamese. This place somewhat resembles an old-time Western fort except that the walls are white masonry and the windows and doors are in green trim. The intent is to have a room for each man, but people come in faster than the building program, and things are crowded. This place is entirely self-contained, with our own power generators, water supply, mess, everything right here in the compound. Trenches, sandbagged walls and walks, and foxholes surround the place, with a machine gun positioned neatly at each corner. Everyone has at least one weapon plus grenades. A rifle platoon from the local Civil Guard is assigned here as permanent protection. Together with the

---

2. *Webster's New World Dictionary of the Vietnam War* (1999), 59–60.

Vietnamese, we muster a lot of firepower, not to mention we're very close to the 40th Infantry and other units of the ARVN 22nd Division.

My job is to run the compound, but unfortunately not to advise. I have a number of officers and men to assist me, while most of the people here live like it's a hotel and go out and advise. I was disappointed at this turn of events because only 3 of us who came here ended up speaking decent Viet, and I hoped for a choice job. But I'm also the youngest and least experienced, and the colonel who is the chief advisor here thinks I don't have enough experience to run a battalion in combat. So I get to run this oversized combat fort instead. Phooey!

Soon after arriving, one of our pilots out of Pleiku, a chief warrant officer, was shot in the groin flying a YH-21 over hostile territory. The ship was almost lost into the jungle, but the co-pilot saved it at the last moment. All our guys were badly shaken by it, as it was our sudden personal introduction to the war. Flak jackets and anything else we could put between us and potential enemy bullets went on chopper floors after that.

Our choppers here *(YH-21s, the "flying bananas")* have been mixing it up in conjunction with the ARVN, trying to bust up a bunch of Viet Cong drifting in this direction. Fighting has been limited to numerous small but sharp skirmishes here and there over this vast expanse of mountainous jungle. This afternoon I went out on a chopper mission with them. Two YH-21 choppers with a machine gun in each door, 4 crew men, and 3 of us armed per chopper moved out. The mission was to determine the combat situation of the 40th and to retrieve an American captain and sergeant who were marooned in a sea of the Cong at Dak Ro Tah. We rolled to the Kontum airstrip at 13:30 hours in a convoy of armed jeeps. At 13:35 we flew over Kontum, above sparsely settled jungle-type hilly country towards the mountains. The machine gunners stood ready in the doors, but things stayed peaceful down below. The roads and trails below were winding with the contours of the land and were mostly gravel or dirt. One main road *(Route 14)* was a narrow strip of tar. This, I believe, was the road to Ankhe [An Khê], the same road where 8 years ago the famous French force unit Group Mobile 100 was annihilated by the Viet Minh. *(Fifty-sixty miles down the road and a left at Pleiku.)*

The villages along our route were not Vietnamese. They were Montagnard. Most do not speak Vietnamese. They do not like the Vietnamese for the same reasons Native Americans did not like the white man. They do not like the Communists, but they do like the Americans and the French, whom they viewed as protectors from the Viets. What a switch!! Their thatched-hut villages dotted the landscape here and there. The roads were almost empty. Occasionally a lone brave vehicle could be seen winding its way along. And I mean *brave*. The Viet Cong has the roads mined and ambushed. The locals can sometimes buy their way through.

The jungle is of two types around here. Neither looks anything like the movie-type jungles *(from my earlier life)*. The first type is a thick type of low growth: short trees that look something like ours, but with every conceivable type of undergrowth to make passage rough. The second type has tall trees that, again, look familiar from the air. The trunks, however, go 150 to 200 feet high before the branches start. When these trees are close together, their foliage forms a thick canopy over the jungle floor, which usually cuts down the amount of undergrowth. When these trees are not so close together as to restrict the sunlight from coming through, the thick undergrowth exists at their base.

All of this presents one hell of a lot of green to look at, with occasional open spaces of red clay earth. Here and there one or two or three huts appear with a small amount of the forest burned and cleared away, as some bold Montagnard family attempts to go it alone (19th-century America?).

Our two helicopters alternated between high and treetop altitudes, wending their way northward toward a 40th Regiment outpost. We fly high to attempt to avoid Viet Cong fire and, at treetop level, to get past them before they can fire, or perhaps to plaster them ourselves.

We watched the ground on both sides of the road for suspicious activity. At this time we saw none. We didn't see any tigers, elephants, or large snakes known to inhabit the region either. *(Actually, elephants were more in the Ban Me Thuot area one hundred miles or so south on the same road.)* We landed at what appeared to be the end of the world. The name of the place I don't recall. It was located on a small mountainside, a 50-man outpost of the Vietnamese 40th Regiment with some Civil Guards *(later to be some of my people)*. The place had a bamboo and barbed wire wall with a huge bunker in the center. Trenches and firing positions surrounded the area. A few huts and tents were in the enclosure. Four 105mm howitzers sat poised in the area.

Two VN captains and a lieutenant jeeped down to meet us. As we entered the area we saw the Vietnamese soldiers improving the defenses.

The captain in command took us to his situation map and briefed us. Forty-five minutes earlier a Ranger unit out in front of him encountered an undetermined number of enemy. Two men were wounded in the fight. The Cong withdrew toward the Laotian border, which is very close. The Rangers were attempting to drive the enemy back into Laos. I batted the breeze a little with the captain in VN, and we found out that they had been hit at this position twice within the last few days. They had four wounded already evacuated. These men were there a long time, sustaining numerous attacks. They were battle weary and suffering from combat fatigue. The Vietnamese captain was asking for help, and there was nowhere for it to come from. I felt sorry for them as we left them in their little isolated bastion.

Now this was the heart and soul of the war from the mid-1950s through the real period of American combat intervention (1965–72) and again until the end in 1975. This on-the-spot report, representing about a year in the mountain highlands in 1962–63, does not jibe with the critics who said the South Vietnamese were not fighting. As far as fighting attitude, I can say, as a lifetime reader of military history, the VN seem pretty similar to anyone else. There were tough times during this long war that would demoralize anyone, especially when a big brother like the US had taken your destiny out of your own hands. The greenhorns copped out like other people's greenhorns copped out in combat, but the hardened veterans by and large could fight as well as anybody else in history. A lot of ignorant remarks were made by certain dubiously qualified Americans during the war, who I think didn't know what they were talking about.

The next leg of our journey was to Dak Ro Tah fort, on top of a mountain, to bring out two isolated Americans. The rainclouds looked ominous *(early monsoon season)* with a storm looking ready to break out any moment. It was a bad time for us because choppers have a hard time flying in the mountains, and they should not fly at all in bad weather *(circa 1962)*.

Down on the road we saw a group of tiny black-clad armed figures heading for Dak Ro Tah. Possible identity Viet Cong? Possibly part of a larger force; we could not be sure. The mountains were becoming steeper and higher. A muddy river wound its way in the steep-ridged valley below. Approaching what looked like a fairytale castle was Dak Ro Tah, and I thought sure old Hall was going to smash that chopper right into a ledge. They finally landed those two whirley birds on the top of a very small but flat pinnacle.

The Montagnard fortress village of Dak Ro Tah was a sight to behold. It was way above us on a high ledge sticking out of the mountain. It reminded me of the legendary fairytale castle. The wood-and-thatch village and its defenses seemed perched very precariously at the top. A dangerous-looking winding road ran up to the village. Our arrival caused a long antlike line of folk to come streaming down to see our whirley birds, Viet Cong or no Viet Cong. They are a very primitive type people, wear little clothing, and resemble the American Indian far more than they resemble the Viets. The men carried a nasty-looking type of spear, and except for the kids (kids are friendly to GIs anywhere), nobody looked real friendly. Our task was to unload, load up, and scram.

Right about this time the weather broke and it started pouring. We were grounded until it stopped. Our two Americans, a captain and a sergeant, were mighty glad to see us. We were glad to see them too because they had lots of firepower and we were not sure how long we were going to be stuck there. We were all thinking about that suspicious-looking column of black-clad figures we had seen coming our way a little way back, and we started calculating when they might arrive. Maybe 45 minutes. We could just see the headlines:

CONG TAKES TWO U.S. COPTERS ON THE GROUND!

We expected the storm to blow over so we could pull out at any moment. Now, generally you have a problem getting American GIs to develop a warlike attitude and dig in, etc. But in this case the first clown to start talking it up and move to a good defensive position was a PFC (private first class) crewman. I was just a newcomer, so I watched him load and reload his weapon, grab all the grenades he could carry, just about stick a knife between his teeth, and expertly find himself a good firing position. Talk about comic relief, that was it. I could just visualize the same guy six months ago in basic training trying to sneak out of such tasks.

Comic relief did not make our situation any better, however, as the Vietnamese garrison scrambled back up the mountain to their positions. We painfully became aware of the fact that we were too far away from the Vietnamese for their fire to cover us. If there was going to be a fight, we would each have to fight our battles alone. The rain kept pouring and a heavy fog settled in with it. Finally the colonel (Tencza), who is a tough combat type, hollered, "The hell with it! We fly out!" It was the lesser of two damned evils, so with some misgivings we clambered aboard the choppers and roared blind as bats out of it. But the day was not over.

By this time, I lost all track of where we were going (like a knothead) because I was covering the ground like a hawk. But we landed at one more outpost to visit our boys there. We had flown out of the weather, and the place looked much safer on relatively flat ground with a big garrison. Our guys and a Vietnamese major ran out of the defensive works to greet us. I thought, "Well! It must be over for today!" Until the Viet major says to the Old Man (Tencza), "Excuse me please, Colonel, sir, but you must please depart. The enemy is right over there and I have only a small squad between you and them." If you could have seen our faces in that moment, it would have been worth a hearty laugh, I'm sure.

I fully expected, and was hoping for, a mad dash to the choppers, but I wasn't going to be the first one to move. The Old Man stood there just like if the Viet major had asked him if he'd heard the score of the Cardinals ball game. So we walked casually back to choppers (the pilots didn't know) and took off leisurely. I guess the Cong must have been at the Cardinals ball game; if they shot at us, I sure didn't know it. About this time, somebody must have told the pilots because they opened those babies up full throttle and flat moved out.

We saw some truly touching things on that trip besides those already related. We flew over one Montagnard village, deep in Cong territory, where the villagers had constructed a huge cross on the ground. Whether they did so as an act of faith or just to keep our air cover from clobbering their village, I don't know, but it was a brave and defiant thing to do.

We once flew over a lone bus braving its way down the winding jungle road in the hopes of safely reaching its destination (probably Kontum), and the occupants leaned out of the windows and waved and grinned as we skimmed overhead. Everyone knew if the Cong saw that, and if it was their territory, these poor people had it. But they gave us a cheer anyway. The guys in our chopper looked at one another and shook their heads, a little bit shook. All along the way people here and there would wave at us; it was quite something to see. And we could see it because we were dodging the treetops. I thought sure Hall was going to catch our fixed landing gear in the trees, but he flew that chopper like a Saigon taxi driver dodging in and out of clearings and trees, making a beeline for home.

When we got in I checked the defenses of our relatively secure area, ate supper, and here I am writing you. Henry, who was also with us, said, "I wish about 10 million other Americans could see what we saw today."

Tonight the monsoon hit Kontum. The rain comes pouring down in torrential buckets. The trenches are flooded, the roof she is springing leaks, the bugs they are everywhere, water is swirling about on some of our floors, we have screens instead of windows and it is spraying in. They say the snakes start coming in from the jungle, so we may be shooting snakes to boot. When the monsoon hits, no planes. So our only contact with the outside is radio, and the telephone when it's not washed out. The roads are either washed out or Cong infested, so here we are caught in between the few planes they can sneak in to us on occasion. The detachment stocks up on supplies every year before the monsoon, so they say we don't hurt for much.

The monsoon washed out our field telephone line to the 22nd ARVN Division switchboard this evening, and a Viet signal team came in to fix it. Instead of just repairing the wire, they have some kind of a funny idea to just put in a new field telephone. They don't take the old phone out, they just put in a new one beside the old one and hope for the best with some kind of Vietnamese incantation. After this Vietnamese corporal and his crew had lined up about 3 new phones on my desk without success, I decided to discover what was going on. So I started haggling with the corporal in Vietnamese and he kept telling me, "The phone is out, the wire is broken, we have to fix it." I kept asking why put in a new phone, just go out and fix the wire. The Viets brought up a 1st lt. to back up their side, and when it ended I had a bunch of new phones, each complete with wire laid all over my doggone office and the old phone sitting there still out. Maybe I should have studied Russian at the language school, or just what do these guys think they have to do whenever a phone goes out? I guess I'll have to take care of it in the morning. The Viets are going to learn better than to leave a bunch of telephones around an American who needs this type stuff. But on second thought, maybe that guy's captain is going to bum him a good one for the disappearance of all those phones. Oh well, tomorrow is another day.

Tomorrow, 19 May 1962

Well, I found out my VNM [field telephone line?] is OK; that's the way
these guys think they have to operate. The phone in the colonel's quarters
went out today, and we had a jewel of a time getting it fixed. These guys
don't tag their wires so you can tell which goes where. They figure it out
by ripping up the wire and trial and error. Gad, what a system! They fixed
the phone and loused it up again all afternoon. We had a livin ball.

When we took supplies up Route 14 by vehicle, it was often a three-
quarter-ton or "deuce and a half" *(two and one-half ton)* truck preceded
by a jeep that ran the gauntlet up the road to Tan Canh [Tân Cảnh]
with a load of supplies and mail for our advisor teams. I frequently rode
in the jeep, "shotgun" with a submachine gun. At that point in time the
road was interdicted by Viet Cong guerillas who were not veteran troops.
I guessed rightly that they didn't want a fight with any regulars who were
alert and ready for a fight. When ambushes happened, they happened
to Vietnamese troops who were not vigilant and were probably sleeping in their vehicles. Their sleep was extended forever in a bloody mess.
Sometimes I would stand beside the road at the gate of the base camp
and pound with my swagger stick on the side of deuce-and-a-halfs rolling
north, exhorting them in Vietnamese to be vigilant and stay alive.

21 May 1962

I arrived in Kontum. I believe on that day they opened up several new
APO *(Army Post Office)* addresses in Vietnam. Then they got the mail
fouled up among these. In addition to being isolated in about every way
you can think of, the mail situation doesn't help. Especially when we happen to know that the mail people down in comfortable ole Saigon have
things pretty soft and ain't about to work one minute overtime to get our
mail out here. To top it off, the knotheads will send our mail by foulup
somewhere else on occasion and send us someone else's mail. Then everybody has to wait several days for a plane to send the mail back to Saigon,
where the lengthy process of getting it sorted out and sent out all over
again begins once again.

The same nonsense takes place with supplies and personal baggage.
I lost a bag full of field equipment somewhere in Vietnam because they
almost put me on the wrong plane. The political wheels down in Saigon
first decided to send me into II Corps area in the southern Delta. Then

they thought I was needed more at Ban Me Thuot and they notified the Vietnamese air force to get me on the next plane up there. Then they finally gave me orders for Kontum but failed to tell the transportation people about change 3. I had myself all loaded on the damn plane when I decided to double-check on everything. (Don't take anything for granted in the army.) I asked a USAF sergeant, "What's the flight itinerary?" He says, "Dalat [Đà Lạt] to Ban Me Thuot." I arched an eyebrow and asked, "When do we hit Kontum?" Says he, "Kontum? This plane isn't going to Kontum!" After two minutes of heated conversation I grabbed my bags off that plane and snapped, "All right, what the hell happens now?" I was directed to Flight Operations, and after several minutes of wrangling over the phone with a Saigon staff officer I headed out to retrieve my baggage. Lo and behold, some overzealous worker had thrown my bag full of field equipment back on the plane and the plane was gone. This little anecdote seems typical of much of the operations going on over here.

Phenomenal progress is being made here in a most difficult rapid buildup. But the myriad of little things where there is no excuse for fouling up are being fouled up beyond belief. I command a very strange situation for the nuclear age. It resembles a modernized version of a frontier fort more than anything else. In another week or so, when it's finished, I will also have a slightly smaller outpost up famous (or infamous) Route 14 known as Tan Canh. That one will be about 40 km away. The road between here and Tan Canh is a ?. It could be anybody's at any time. We can still get small armed convoys through, but believe me it is only because the Cong has made no real effort on that stretch of the road. A few sporadic shots and last week a civilian bus hit a mine and blew up, but that's all. But the Cong have enough troops to close the road with some real action any time now. Our problem is figuring out when?

I have off and on 100 Americans around here. They go out and work with ARVN units around here. I also have a 50-man platoon from the 40th ARVN Infantry as a security guard and numerous Viet troops for various housekeeping details. This place is laid out, as I said, much like an old frontier-type fort. Each corner and various key spots have a machine gun post, and each American and Vietnamese has an assigned defense post. Trenches, gun emplacements, sandbags, logs, and earthworks with barbed wire and bamboo spikes surround "Frank's Fortress," which we might otherwise call the Padarewski Brigade because we are loaded with Polacks.

If you have seen pictures of the typical fort over here you might ask, do I have a watchtower? Yes, we have a concrete combination watchtower/water tower with a couple of ARVN soldiers in it, overlooking the scene here.

One of these days I am going to take a patrol out and scout this area to verify the info I already have about the terrain around here. The people who must know this info in great detail already have it, of course. About 500 meters west of us, just across Route 14, and behind a rather intricate Buddhist tomb, is a gully. That's where they think the Cong might emerge from if they ever do again.

I have to fly back to Saigon for some conferences on 2–3 June. One thing I want to get my hands on for the benefit of that gully is a couple of mortars, which at present we lack. We have at present, I think, enough firepower to require a Cong battalion *(consisting of six hundred to eight hundred men)* or more to carry this place. In addition, ARVN 40th Regiment and 22nd Division units are within a very safe distance. The Cong doesn't appear to have enough units about at the moment to try us out. The task does interest them, however, since an ARVN patrol surprised a VC recon party up the road a few weeks back and captured documents assigning this group the mission of sniffing us out. At the moment they seem a little confused about my defenses here.

Presently the main Cong effort in this area seems to be a great influx of troops and supplies from the Laotian border north of us, across the central mountain cordillera to the coast, where both friendly and enemy activities are building up for a coming showdown of sorts. Activity in our area by the enemy is sporadic just now, but plenty dangerous enough for us. The main efforts of ARVN here is to attempt, with its limited means, to pummel as many of the enemy as possible, and at least to push them back into Laos. The better the job our local ARVN boys do, the more likely we are to stir up action around here. The enemy, who may also be short of people, may attempt action around here to keep ARVN busy. Then their routes of communication across the mountains would be less harassed.

ARVN is much more active under US guidance than before, under French influence. We try to push an offensive policy; they learned defensive stuff from the French. But this is a long, hard go. In this area we are partway there. But there are still too many units tied down in forts. I saw a few such outposts myself the other day. There is a good reason for a place like mine here. Even if I would rather be chasin' than waitin'. This place is nothing more in theory than a base for offensive-minded training and operations of our many advisory teams here. Small teams of my guys leave here continuously to go into the field for a day, a few days, months. We set up Tan Canh outpost to move some of them even closer. I have a rough task maintaining these guys. When they can't take what they need along, we have to get it out there to them. Sometimes this means a risky convoy. Usually we take it in by aircraft, such as helicopters.

If our tactics and strategy work and ARVN does a real job on the Cong up here, the US MAAG is the principal reason. The logical thing for the Cong to do is start after us in earnest. There are possibly 4 reasons why the Cong is not after MAAG bases like mine at this time:

- It involves the highest-level policymaker's decision, and they are very queasy about starting anything more with us *(the US)*.
- They don't want to divert the effort from their main operations at present.
- We may not be making ourselves felt enough yet to justify their going after us.
- Maybe we're making ourselves felt so hard that the Cong just doesn't possess the resources now to take us on too.

These are thoughts that naturally revolve around in my head from day to day, and whenever I walk around my outer perimeter and gaze out on the knee-high undergrowth around us, I always wonder who might be looking back.

I made a very interesting helicopter flight the other day with Col. Tencza, who is my boss and the senior advisor to 22nd Division here. I cover that elsewhere. The next day the Old Man went out again and they shot up the copter. Nobody got hurt, thank God, but they just missed the Old Man. The Cong is getting good on shooting at choppers, as you've probably read. Our choppers go armed with a .30 caliber machine gun each in the forward and aft doors. In addition, everybody else on board goes well armed. They drive the Cong nuts, and the Cong has issued all sorts of special instructions and incentives for shooting down choppers. The chopper boys get shot at fairly regularly around here. We generally have several out at the airfield here, with crews staying within the walls here.

Meeting incoming aircraft around here is a comic ritual. The airstrip is a steel mat one-horse job guarded by ARVN. It has one tiny little Air Vietnam building and a passenger waiting stand that is like little more than a corner bus stand back home. Air is virtually our only contact with the outside, and mail, supplies, etc. come in on the weekly "milk runs" or by some unexpected flight. Incoming aircraft depend on us for transportation to the compound, and we depend on them for news, food, and ammo. The signal for the race to begin is the sound of an aircraft overhead. The pilots play the game by buzzing the compound on their way in. This starts a mad dash of soldiers strapping on weapons, piling onto every conceivable type of vehicle, roaring out in a great scramble of honking horns, waving weapons, and gunning engines. Out the gate we go, leaving behind the unlucky ones and the ARVN guard gawking at us. Roaring down the Kontum streets we go and into the mighty metropolis airport, "Kontum International." As the aircraft *(usually a USAF C-123)* taxis in, the crew is greeted by a great cloud of dust coming up the road, which materializes into 5 or 6 different vehicles storming across the field in disarray with wild-looking inhabitants (like Pancho Villa's band), waving weapons, hats, etc. By the time the crew climbs out, the vehicles are neatly lined up, everyone is silent, all eyes look like eager puppy dogs', and maybe one meek voice will ask, "Did you bring any mail?" We can't eat mail, we can't shoot mail, mail won't run our vehicles, but when the clowns down in Saigon messed it up, we found out the mail can run *us*. But we all know everybody writes to us and someday we'll get it, and that means a lot. Some of these guys were members of the original crew of 9, and they are counting the days in ceremonial ritual until they can return home. Spirits have been and are good.

Well, it's time for me to hit the sack. Many, many problems confront me in this operation, which is suffering so much from growing pains. Tomorrow I have to muster my cohorts and move out sharply. There are only two of us who speak Vietnamese. One's a sergeant. I'm the other one, and I speak it the best—which is fair. The Viets around here didn't get as excited as the Saigon people at first, but word got around and they seem to know me all around town now. "Oh yes (in Viet), so you are the American captain who speaks Vietnamese very well." Well, I don't speak it very well in reality. But surprisingly it looks as though I will very soon. I've got all kinds of offers to help me finish learning it, perhaps in exchange for a little English. My best offer yet is from the vice governor of the province here. (Maybe I can hold out for the governor himself?) You ought to see the absolute switch in attitude of these people toward you when you speak *their* language, even only fairly. The more different subjects you talk about, the more impressed they get.

The crowning event occurred this morning in the vice governor's office. There was a vital exchange of information going on between the vice governor and Major Brady. The interpreter was having a rough time, and the tactical situation on the map in Viet was obviously not getting across to Major Brady. I hadn't said boo up to then, so I had a damn ball stepping up to the map, clarifying the situation in Viet with His Excellency, whose eyes about popped out of their sockets, and then turning to Major Brady and explaining in English. For a dumbhead as linguist, that was plenty of fun.

This sort of thing has also backfired, however. I also find myself now telling the Viet work details how to scrub down the damn toilet bowls American style. C'est la guerre! I also discovered when there's something wrong that they don't want me to know about, these Viets suddenly talk a mile a minute. But when they speed up they get that guilty look—you know? Hah! Catch 'em every time. More anecdotes from this isolated bastion of democracy (Joseph Alsop!?!) later. Must be 300 and fifty some days until I come home. Don't let 'em build the tailfins too big on the cars back there.

Kontum, 26 May 1962

We are busy here. We move at a rough pace. We have much to do to make the place both safe and livable for our guys who are out with the ARVN. My boss, the colonel, is a terror but a good guy at heart. "Terrible Tony" is his army-wide nickname, and I can vouch for it. This guy would just as soon tear you apart as look at you, and he can do it as well or better than some of the hot rods I've served under. He's a paratrooper type who has lived in the paratroops all his life. (It was said he made the D-Day jump in Normandy and a combat jump in Korea.) Even the Viets jump when he growls, and they live in mortal fear of his anger.

I'm used to handling this type (yeah!). He keeps me on my toes, but I don't think I'll lose any sleep over it. He can be extremely fatherly at times. He called me "Mike" from the start. He likes to have private discussions about this thing or that, and he treats you like a son. But when something looks wrong—LOOK OUT! All hell breaks loose. He put a West Point lieutenant and three or four other captains under me. He roars, "Work that lieutenant to death! Make an officer out of him!" Gad, what a routine this guy sets up. He thinks of everything a step ahead of me and then wants it done yesterday. I think I am proceeding to do a good job here, and he told me once, "I like the way you handle yourself, Mike!!" Then the next day he tore us apart from one end to the other and had the pieces for breakfast. But you know where you stand with this guy (yeah!). There is no bureaucratic guile and hanky-panky. You know what he wants, and even though he is a fearful driver, all he really wants is reasonable progress. He knows how to get it. My goal is to get a step ahead of him. That's a proposition that makes the time go fast. Lord help the Viet Cong if they attack this place while he is in it! There really isn't a lot of time for relaxation. With or without "Terrible Tony," there's plenty to do.

Down at Pleiku where ARVN II Corps HQ sits, we have another terror of a colonel who is also a paratrooper and is "Terrible Tony's" boss. This guy's name is "Coal Bin Willie" Wilson. A frightful terror known to all paratroopers, he has even "Terrible Tony" scared. It's no damn wonder the Cong has been quiet around here recently. Gad, these guys are fierce! These guys would just as soon tear a hunk out of you as look at you. The Vietnamese troops in this region were *(or were said to be)* the best they have, and it's no wonder. They are more scared of "Coal Bin Willie" and "Terrible Tony" than they are of the Cong. Oooh, how they hate to face these guys when they goof up. I've seen more than one Viet officer just plain cringe in terror before these guys. You have to see this sort of thing to believe it. (I wonder what we Americans looked like.)

In retrospect, this is another major area where our government failed in the early war. Personal experience and historical evidence seem to confirm that we did indeed place some of the best officers we had into key spots in Vietnam. If we'd seized the moment, we could have been intelligently turned loose on the situation using the Principles of War [see the Appendix], and we could have accomplished something. But the Kennedy administration's technique of running the government with a handful of "elite" advisors to the president who did all the thinking for the country did us in. Private cabals everywhere mucked us up. Instead of just reading someone's memoirs, if one also reads the now declassified official documents, it becomes clear. A person like myself, a veteran, or a historian can find supporting evidence now to refute the revisionist's history. Those of us who attempted to appeal policy back in the mid-sixties found our military careers had to be sacrificed, or compromised, or we had to try to work it out within the system, which to me didn't look possible. Unlike some of our leaders who kowtowed, some of us didn't, in the hopes of protecting the nation from what was to come. Some of us were badly treated by some people as the months and years progressed, even up into the 1980s.

Kontum has several Catholic churches run by the French and Vietnamese. There is one American Protestant mission here, and one hospital, truly a Tom Dooley[3] type affair run by three heroic middle-aged American women, a Dr. Smith and two nurses. (I think at least one of them was from Milwaukee.) They care mostly for Montagnards, but also for emergency Vietnamese cases. The Viets have their own medical facilities, such as they are. They [the Americans] run this operation (the hospital) on a shoestring. The conditions they work under are unbelievable. When I visit I can hardly stand to be in the place where they are working. These gals go places where we don't dare go. All of us Yanks are pretty much in awe of what they are doing.

The April 1962 issue of the Catholic magazine The Sign *published a report by Dr. Patricia Smith, a thirty-five-year-old American doctor from the Catholic Relief Service.[4] She lived with two nurses from the Grail, another Catholic relief organization. As she wrote, "I don't know how many people I have watched die because I didn't have a bottle of blood to give them. And we are always running out of antibiotics."*

Smith first arrived in Kontum in July 1959. The only medical facility was a leprosarium for 250 patients, headed by Sister Marie Louise, a brave French nun of the Order of St. Vincent de Paul. Smith described the work:

> *Gradually the Montagnard began coming to me for treatment. Their numbers increased so much that they filled the house, and I was treating fifty people a day. Leprosy is a terrific problem, but the new problems I was meeting were bigger and more demanding. The general medical condition of the people is unbelievable.*

3. Dr. Tom Dooley, a household name at the time, had become famous for sacrificing himself to minister to the Vietnamese people in the fifties.

4. Patricia Smith, "Doctor Smith and 300,000 Tribesmen," *The Sign*, April 1962.

Everybody to begin with has malaria. Their diet is one of the poorest I have ever heard of. It consists of a little rice, some ferns and bamboo shoots and green leaves, and a few small animals like rats and mice, lizards, snakes and this kind of thing. (That's what I ate as an advisor when in the field with them, and also, I believe, insects.) *Intestinal parasites, worms and hookworm are a big problem in our area. All this weakens their resistance. In addition to 10 per cent of the people having leprosy, another 40 per cent have tuberculosis. The climate is mild, but the rainy season brings on pulmonary diseases.*

Three quarters of the children die before they grow up. Actually a large number die within two weeks of birth, because of the terribly dirty way deliveries are accomplished. The women for the most part deliver themselves, or they have the help of an old granny in the tribe who has a reputation, though no training, as a midwife They have no idea of germ theory. The woman is back in the fields in a day or so. Many children die as soon as the mother runs out of milk. Then more die at the age of four or five from a variety of diseases which they don't have the resistance to combat—pneumonia, dysentery, malaria. And again, at the age of puberty, many girls die. They don't have that extra something it takes to meet this period of rapid growth.

Another reason that some of the babies do not survive is that they are murdered. In the Jarai tribe, a wedding is always accompanied by a huge orgy. The father [meaning the groom] is then expected to kill his firstborn child, the reason being that he can't be sure who is the father. They have no idea of time, so the father kills the child, even though it might be born a couple of years after the man and woman are married. In spite of this, I have never seen a tribe that likes children better.

With conditions among the Montagnard as I have described them, it is not modesty but truth to say that we have made only the smallest of dents in improving the health of the people, even though we have saved many lives.

I remember a sixteen-year-old girl who was suddenly struck with a strange and prevalent form of cholera that can kill a healthy person in four hours. "The girl was in shock when I got to her," said Dr. Smith [the author is writing about herself in the third person].

Fortunately we had just received a big shipment of intravenous flu-ids from the United States, so I hustled back to the dispensary to get some. When I returned to the girl's village, I noticed some people working on a log underneath the house.

"What are you doing?" asked Dr. Smith.

"Making a coffin," one of the men answered. "The girl will die." And he went on hollowing out the tree trunk. But I got the fluids into the girl in time, and the next day she was better. The people thought I was a miracle worker.

. . . When it gets too crowded (in the dispensary) *we move all the beds outside and put the patients on the floor. At that, we keep only the ones who simply can't walk. We have no X-ray equipment and can do only very minor surgery.*[5]

This article was published a few days before I arrived and became acquainted with this work. These gals at the hospital were dearly loved by the Vietnamese and Montagnard people they took care of from the region, and the Communists hated them and threatened to destroy them. They didn't get to do it while I was there, but after I left, the enemy shot the hospital up and eventually the missionaries had to leave.

There is a French Catholic bishop here. This is the diocese of Kontum. The priests here go out to the Montagnard villages, Viet Cong or not. A few weeks ago one of these priests was martyred by the Viet Cong. Another was promised the same fate if he came back. He has continued to go out to the villages. Our soldiers here, including myself, we watch this pathetic *(my opinion at the time)*, heroic, and cheerful beloved char-acter of a holy priest *(in rags, with a walking staff)*, knowing each time he may go out may be the last time we will see his cheerful face.

(The Protestants—I believe a Christian Missionary and Alliance husband-and-wife team also named Smith—appeared to be in their late twenties or early thirties. They came by and fellowshipped and ate with us from time to time.[6]*)*

The colonel, "Terrible Tony," who is a Catholic, has a very soft spot for these heroic people [the church workers]. He sees to it that we do all we

---

5. *Smith*, "Doctor Smith."

6. Editor's note: This couple was likely the Rev. Gordon and Laura Smith, an American couple who served the CM&A church in Vietnam and Cambodia from 1929 to 1973.

can for these people, but it seems so little. "Terrible Tony" told me to do all I could, to give them whatever I could. This hasn't been much as yet, comparatively speaking for the lot of us. When you walk into a situation like this you feel so humbled that you feel helpless. *(As a steward of the US Army's war property and the mess fund, I couldn't presumptuously take from these and give it away. I had to look for valid and honest opportunities. And we military were not exactly rich men on our army pay. It seems unbelievable now, years later, but we had to pay for our own support in Vietnam through the mess fund, and many of us also had to support families at home. We did get a pay allowance that helped.)*

We have had these folks over for dinner frequently, and we've taken food over to them, and lent them our vehicles. The Montagnards crowd into the hospital from miles and miles from the mountains to be treated. The cases that you can see when you go over to their little clinic are enough to dumbfound you and turn your stomach. The "clinic" they work in is nothing more than a dirty old house with dirty old slummy furniture. The mountain savages in their stench and filth crowd around the (open) doors and windows and in the "treatment room" itself with their unsightly wounds and diseases. It is a losing battle to keep anything clean around here as it is, but these gals looked like a group of plumbers after ministering to these people, not like the sterilized medical types you see back home.

To try and make a little money to buy more medicine for the ladies, we buy crossbows from the Montagnards and sell them amongst the numerous souvenir-hunting types of American VIPs that come through here. The crossbows are powerful wooden weapons that can bury an arrow 3 inches into a tree at 50 yards with fair accuracy. They sell for 1000 *($1.25)* each from the mountain men, and we try to push them for 3000 and give the proceeds to the clinic. This is "Terrible Tony's" pet project and one of my numerous assignments. I'm in the thinking stage of trying to come up with some real help for these gals. As a first step, I think we will give as a family all of our 10% right here. I'll take care of that. From what I can see, it looks as though 10% from a family with an income such as ours can do a great deal here, comparatively speaking. They find it difficult to purchase the most basis medical supplies that they need. There just isn't enough of anything—especially time. I tell you it takes guts just to walk in the front door of that place. Our military compound is a paradise in comparison. Somehow, someway, I'm going to put our resources together here and go in there and help out.

Kontum, 27 May 1962

Arose 0630 this morning (Sunday) and made 0700 Mass at a school for Montagnard boys. Kontum is a beautiful place. The mountains rose above the fog and the early morning briskness is everywhere. Even the heat loses its edge in the early morning. We have not had a serious rain in a day or two and the humidity is bearable. The school buildings and the church are picturesque. The interior of the church was very beautiful. There were no pews, only a few kneelers for the few visitors such as us. It was kneel and stand on the floor for the Montagnard boys, who were lined up in stair-step fashion with the little ones in front. They wore their colorful American Indian–like costumes with robes over their shoulders. An old organ played by a Viet put out very good music, and the boys popped to the sound of the clapper banged together by a Vietnamese seminarian. A French priest with glasses, tapered beard, and a fine tenor voice, a Frenchman if I ever saw one, sang the High Mass. One of the boys chanted the Gospel in one of the Montagnard dialects. To tell the truth it was one of the most beautiful Masses that I have ever seen. The singing was magnificent. I didn't expect to see anything like this in Kontum, Vietnam. General Harkins is visiting my fort tomorrow. Should be interesting.

The Christmas story out here last Christmas was a corker. There were not too many people here then; "Terrible Tony" wasn't here yet, and there were only a couple of buildings. The mail quit coming about two weeks before Christmas, and each time a supply plane came in the two sergeants responsible for meeting it came back with no mail. As Christmas drew nigh a few more individuals found excuses to join the sergeants to meet planes, but still no Christmas mail. Christmas Eve came and there was no mail, not even a scrap of paper from home. Gloom was everywhere as the boys cried in their beer. Christmas morning arrived, but it wasn't much of a Christmas. It was just another workday because there was nothing to celebrate it with.

Suddenly, someone heard the distant drone of a plane. It grew louder. No one made a move. Everyone looked up to the mountain pass that planes come through when they come here. Someone saw a speck. The speck grew larger. It had to be mail at last! With a wild roar the whole damned outfit jumped on anything that had wheels on it. Kontum had just recently been attacked by the Viet Cong and the compound had got caught up in it, but even the guys on alert took off. Only a few Viets were left behind. The whole outfit dashed pell-mell down the streets out to the rickety airfield and waited like a bunch of kids, field grade officers, captains, grizzled old sergeants, and young privates alike. The plane landed and taxied up to this pathetic little group full of the yuletide spirit and stopped. The door opened. And who do you think stepped out? Santa Claus? IT WAS "COAL BIN WILLIE," THE SUPER TERROR! And he was mad as a hornet because the whole outfit abandoned the fort and came to the airfield. He stomped and roared and tore everybody apart. How would you like to be hoping for Santa Claus and get the Big Bad Wolf instead? When it was over the colonel got back on his broomstick—er, plane—and flew off. I guess the boys got a good laugh out of it. It's one of the local legends around here. (Since the place was recently attacked, one can understand his reaction a bit.)

Kontum, 29 May 1962

Outside of church I have not heard any music since I left. I have long since run out of pipe tobacco; even that stuff I mailed to myself two and a half months ago has not arrived. Besides that the bugs around here are a real nuisance. There are some huge types which you don't see real often, but often enough to scare the hell out of you. There are some types of beetles that go up to four inches long and well over an inch fat. There are some monster spiders that I've seen half as big as my hand. We seriously consider whether to shoot the critters or spray them with DDT and club them down. Can't shoot for fear of ricocheting bullets and setting off a general alarm. But DDT American style, such as Off, is for some reason useless over here. The army issues a bug bomb that slows em down enough to clobber them. I keep a big club to chase bugs with.

I had a hell of a fight the other day with the biggest spider I ever saw. [Before going to bed for the night] I took the bug bomb and sprayed under the door just for good measure. I leaned over and let er rip. A monster spider leaped out from under the door at me and just missed. (No wonder I just got pasted to the roof!) I don't remember how I got there, but the next thing I remember I was a good six feet away on the sleeping bag, waving my pistol. I didn't touch the floor getting there, either. That big brown-and-black spider stood there in the middle of the floor looking up at me, and I stood on the sleeping bag looking down at him. I decided right then and there that both of us were not going to share that room that night. When he realized that, he shot across the floor behind my footlocker and I ran out and grabbed the biggest club I could find. When I came back he was peering around the corner of the footlocker at me. Back and forth we went. I stayed my distance and swung the club. The huge monster kept jumping out of the way. I stayed far enough away so he couldn't jump on me. Finally he ran behind the footlocker. I decided the bug bomb might drunk him up enough so I could hit him. I emptied the Off can behind the footlocker and, sure enough, that monster staggered out like a drunken sailor. WHAM!!! It took two swats, but I got him. This sort of thing happens frequently around here. *(There was an even bigger bug in an episode I wrote about later.)* Sgt. Uremovich (Mark for short) chased one around with a Montagnard spear until he finally harpooned it. Man, what terrible-looking creatures!

The monsoon keeps looking like it's coming, and then it won't. Humidity is so bad enough, even in sunlight, that it keeps every stitch of cloth damp. When the sun is out from noon to 3 p.m. it's exhausting to keep going. *(The Vietnamese took a siesta every day at that time. Many of us had trouble doing that because of our North American culture and work ethic.)* You spark up a little in the evenings, though. Sometimes it gets less warm enough that I use a sheet for covers, but everything is damp.

Some people from the US 27th Infantry Wolfhounds in Thailand are passing through here for a look-see at the war here. If they stick around for a few weeks when it gets too wet for the choppers to get off the ground, they might see something.

The town of Ban Me Thuot, which was mentioned in a press article, is also on Route 14 south of Pleiku, which in turn is just south of Kontum. The whole region is known for the fighting that took place while the French were here and, of course, recent fighting.

We don't get the news. We get newspapers (*Stars and Stripes*) a week or two old. There is no radio except for our fighting-type radios. Sometimes news comes in like in the Old West, by plane or convoy. We hear rumors that we put another man in space. Is it true?

I put Doc Smith, her truck, and a sick French nun on one of our planes for Saigon this morning. She's a corker. I sent a truck, a jeep, and 20 soldiers armed to the teeth out into the bush to help her out Sunday.

She and those French priests around here go out completely unarmed, even after the Cong devils martyred one of the priests a few weeks ago.

Hope to lead a motor recon down to Pleiku to get the lay of the land around here. If you have more firepower than they have, they don't fool with you. A lot of shooting generally brings help fast, so all in all, no sweat. The Cong have been rather unskilled in what little trouble they have pulled on the roads around here *(at least at that time and place).*

Kontum, 3 June 1962

There was an article in the *Saturday Evening Post* in December 1961 or January 1962 titled "Last Chance for Vietnam."[7] Mention was made of the outpost of Khon Brai [Kon Braih]. Enemy activity up there is stepping up considerably. Route 14 (the French-built highway, blacktop part of the way, dirt the rest) has been mined and ambushed 5 or 6 kilometers north of here. Two days ago a Viet squad (about nine men) got hit on the road. Armored cars and half-tracks roared past here to their relief. Khon Brai and other small outposts appear threatened once again. This afternoon I went out on a helicopter supply run to Khon Brai. The other chopper landed and delivered supplies while ours flew cover as gunship overhead. I took some good movies. Didn't see anything threatening.

Saigon, 6 June 1962

The climate seems to aggravate whatever ails most guys, and sickness has been a problem compared to our previous experience elsewhere. The battle for sanitation and proper diet is constant, one that bothers us at the moment far more than the enemy. *(Studies in the years since then have revealed that this comment has been the comment of the ages for the entire history of the planet. Only in the post–World War I period have armies been able to get on top of these issues, if at all. I did not properly appreciate this at the time.)* It is extremely difficult to keep *anything* clean. When it's not raining, dust gets everywhere; when it rains, mud gets everywhere. In this climate it doesn't take long before the best-constructed buildings (for Vietnam) settle, warp, crack, and start to deteriorate under the elements.

This creates places where bugs, rats, and reptiles can get in and add their pleasant presence to life. Drinking water is hard to come by because it must be strained, boiled, and chlorine added. The rest of the water we

---

7. Don A. Schanche, "Last Chance for Vietnam," *Saturday Evening Post*, January 6, 1962.

use for washing, etc., is considered highly contaminated. We have some hot water heaters at Kontum, but they are too small and break down constantly. The plumbing is very poor, the Viets can't seem to plan, install, and maintain a really adequate plumbing system. The job is made more difficult by the junk that just naturally gets into the pipes in no time at all due to the climate. The chemicals we Americans use to clean pipes are in short supply in Vietnam. Plumbing in Kontum is just one of my problems, but no small one. Without adequate plumbing we face serious sanitation problems daily. Nothing that we have is easy to clean. We have to boil and specially treat water to clean dishes in. We have to have special chemically treated water to wash off vegetables with. Buildings, furniture, latrines, and fixtures are constructed with difficult materials to clean, such as wood, plaster, or concrete. To top it off, the Vietnamese detailed to work for us are another potential source of sanitation problems. Life itself is difficult to maintain over here. *(Despite my comments at the time, I don't think either I or my comrades realized how good the American system of field sanitation and nutrition was, and how recently developed it was, considering world history. I did that job well because I was well trained and disciplined and completely overlooked its importance vis-à-vis the combat stuff. We did not have the "sick call" of previous wars in history in similar and much better situations because our army took care of us in this area better than any other in history. Informed historians can appreciate this. See also* From Sumer to Rome: The Military Capabilities of Ancient Armies *(1991), by Richard A. Gabriel and Karen S. Metz.)*

Communications with the outside world are another problem. We keep a couple of helicopters *(H-21s)* at the compound. Several supply planes come in each week at the airstrip. Once in a while an armored convoy gets through. There is a telephone line to Pleiku that gets cut almost daily. Road travel is restricted to armed convoys. We have good military radio communications within the compound. We get the *Stars and Stripes* newspaper about a week late. We don't know much about what is going on outside. We still have a couple of stateside papers that someone received in the lounge. Everyone still reads them, though they are dated two and three months back.

Kontum, 8 June 1962

Fighting is sporadic around here but has a definite pattern to it, and the coming months should increase activity on both sides. I provided myself with a series of scares this past week, some real, some imagined. A week ago the Viet Cong suddenly waylaid the roads around here. They've been able to do that at will anyhow, but they were not bothering Americans for some reason. Friday and Saturday last were active. They especially hit the Kontum–Khon Brai road *(Route 14 up north)*. They set out mines and ambushes. One ARVN squad got caught in an ambush up the road a few thousand meters, but our reaction was not good and they got away. Twelve hours later armored cars, half-tracks, and jeeps with machine guns mounted roared up the road, in the best traditions of the French defeat, to find absolutely nothing. I could have told them so, as I watched them roll by my little fort. Saturday morning I sent a small convoy to Pleiku, composed of American and ARVN soldiers. They drew fire, but nothing serious.

Sunday I decided to go along on the helicopter resupply run to a few of my people at the famous outpost of Khon Brai. A beautiful view and day, excepting the war. We flew at treetop level in two copters. We expected a belly full of automatic weapons fire at any moment and were covering the ground in tense anticipation. Nary a thing happened.

Monday I boarded an Air Vietnam C-47 for Saigon and enjoyed a good scare. The passengers included a number of peasants with noisy, smelly livestock. The Vietnamese and French pilots who fly those old crates will fly anything anywhere, with the greatest amount of daredevil spirit. American pilots outside of Vietnam generally avoid rough weather. But the Viets fly like they drive cars, and this guy ploughed and bounced through the early monsoon squalls until I thought we might break up. We flew through mountain passes, over the mountains, and hit numerous air pockets that really threw us around. At one point we plummeted downward several hundred feet near the treetops, and I stayed up in the air, held in by the seatbelt but with the seat handle wrenched off in my sweaty clenched fist, amidst feathers and squawking galore. But lo and behold, he set er down on the Saigon airfield just as neat as you please, and I pulled myself together and walked away from it. *(Anyone wants an airplane story, I've got em from my time in the military.)*

Another scare came this morning when it was still dark and I waited alone in the Saigon streets for my jeep to return me to the airfield. You normally don't do things like that alone, but there was no choice. I curse myself for leaving my little Spanish automatic back in Kontum. The jeep didn't come and didn't come, and a small group of shady-looking characters began to gather around me. They appeared friendly and were greatly impressed that an American captain would make small talk in Vietnamese with them. I couldn't make up my mind what these guys were after, and the jeep didn't come, so I made a gallant goodbye and grabbed a taxi. This guy proceeded to take me to the airfield by a dark and spooky route that I did not know, and I was beginning to picture him counting up his reward money for delivering me to the Cong. My wild but practical imagination was mollified with a great sigh of relief as I was delivered safe into the hands of tough but friendly-looking American and Viet Air Police at the airport gate. I boarded the US Air Force C-123 feeling somewhat better about the whole thing, because I thought, "Here at least is a plane I can bail out of if need be." *(C-123s were used in paratroop jumps.)*

The flight north up to Ban Me Thuot was great. Then came the clouds, the squalls, and the mountains. These planes carry supplies, and they don't turn back if they can help it, because somebody might need those supplies. Approaching Kontum, the pilot, who was taking us through some bouncy stuff, began to search for a hole to go down through. One thing I learned about flying is that when these guys have to come down through soup with instruments, it's strictly go as you see. All flying over here is that way *(circa 1962)*. The plane circles down in a spiral, with the pilot tipping and tilting it every which way to see better. You look up into the cockpit and start flying it right with him. When he starts getting out of his seat and stretching this way and that to see something in cloudy mountains, you start looking too, while everybody acts like this is routine. In fact, everybody is plenty concerned. We all saw the mountain we were diving into at the same time. Gad, it was close!! But the pilot, to whom I tip my hat, bless 'im, calmly righted the ship and we eventually found the pass and the needle in the haystack, the Kontum airstrip. I plainly see why they can't always get the mail and the supplies in now, and what a gallant group they are for the days like today when they come in anyway.

This afternoon I decided to pay a call on my little outpost at Tan Canh up Route 14. So I took two jeeps, Captain Ed Mixon, a fine engineer officer, and several of our trusty American and Viet soldiers armed to the teeth. We put sandbags on the floors of the jeeps to help absorb the impact of mines. Everybody carries automatic weapons cocked and ready and keeps grenades handy. The weapons are always pointed in the most suspicious direction, because these type actions occur in seconds, and a reaction must be immediate and violent. This has the effect of scaring the hell out of the natives we suddenly come upon, because everybody knows that the Cong and the natives look very much alike and there is frequently the rapid mental size-up of the situation and the decision whether to shoot or not. But we sort of have an unwritten law that we'll let them shoot first. The French and the Viets and the Cong are not too particular about shooting first and maybe asking questions later. But I do not relish the idea of shooting up some poor innocent guy who can barely keep his family alive. Thank God we are not trigger happy, and so far not too many incidents have happened to make us that way.

The next step in the procedure is to drive like hell and as fast as possible. This lends itself to the possibility that we can whisk past an ambush before it can set itself to get us. At least our momentum will carry us through it, a cardinal principle of survival: KEEP MOVING! I think this may be one reason, coupled with the present demonstration of poor ambush technique by the Cong in our sector, why we get through. It may be we were rushing past a few poorly conceived ambushes without ever knowing they were there.

If a vehicle up front gets stopped, the succeeding vehicles slow down just enough to pick up whoever can jump on. Those who can't are told in prior briefing to run right through the ambush into the jungle and make their way to the closest friends. To be pinned down in an ambush is flat curtains. Yet man's natural instinct is to hit the dirt and get pinned. I think we are getting our people well enough trained to react properly now.

At any rate, Route 14 turned out to be an ambush haven. The narrow bridges, winding road, hills, thick undergrowth, all lent themselves to many, many perfect ambush sites. By the time we reached friends at Tan Canh my eyes were smarting and I virtually had to pry my grip loose from my automatic weapon. The marks of places where the most recent ambushes took place were still fresh. I did not think the Cong did a good job at those places, and I saw dozens of better sites they could have chosen. But sloppy job or professional job, it's all the same when someone might open up on you at the next turn in the road. The war was plainly on the faces of the people we passed; some were scared, some sullen, some brave souls waved and grinned. Nobody knew what the other was going to do, and we covered everything that moved suspiciously.

I felt every bit as if I might have been a French paratroop captain doing the same thing a few years ago. Now it was my turn, the American paratroop captain. Many thoughts crossed my mind. The country was beautiful. The jungle was not the palm tree type, but more like our forests with thick undergrowth. The hills and mountains and the Montagnard villages made the entire scene strange to me. It seemed strange indeed to sit there tense and expectant during that wild ride through beautiful scenery in such utter kill-or-be-killed circumstances. I saw everything but remained keenly alert and self-confident, as I should have been. (I had been splendidly trained and experienced in preparation over several years for this.) I feel more sure than ever now that I can operate efficiently in war. The great bogeyman that is the Viet Cong has burst his bubble before my eyes. I have physically seen what he has done, and I can see that he did not do these particular things as well as I would have done them. I know we can beat them now, if only we can *catch* the so-and-sos *(one of the big tactical problems that confronted us in that phase of the war).*

There was one emotional experience, though, which was certainly different from maneuvers. I did not feel the security I had felt in training with a weapon in my hands, though I was better armed than ever. I had always felt that the *feel* of my weapon would be one of the little but important crutches that would help sustain me in war. But my experience so far over here is different. Even though I probably will be smarter and better armed and a better leader than my opponent, the psychology of the guerilla favors the enemy. He hits me when he chooses. He does not have to be a hot shot to get me. Perhaps this causes the lack of security that grips us. Soldiers in war look to the strangest little things to sustain their inner feelings of security. Obviously our enemy, the Viet Cong guerilla, is working us over psychologically to destroy all of our little mental bastions of security. They can succeed in shaking us up considerably. But I do not think they can destroy the inner framework that makes us fight them. We are right and they are wrong. I wonder if I see a manifestation of our guys accepting their challenge and playing out the game with them. Because once one of my men makes one of these patrols, he always volunteers in a cocky way for the next one. It might almost be as though our guys are shouting back into the sinister jungle, "All right, you clowns, we'll beat you at your own silly game. You stupid clowns don't think you're going to hoax us with all this mumbo-jumbo, do you?" An interesting thought, but I'm a soldier, not a philosopher. In any case, nothing will happen to me unless Almighty God wills it. And He will take care of me as He has and will all of us, in His own good time. So let life march on as He will have it, and He will take care of the things we can't.

Kontum, 9 June 1962

The monsoon is a daily problem, but it is not fully upon us yet. First it rains a half hour or so in the late afternoon and evening. Then the rain increases every couple of days to earlier in the afternoon and lasts longer. Finally it gets into the mornings and it starts raining continuously without stopping for weeks at a time. Torrential rains hit us during some period every day now. Flying, which can't be avoided in a war, can be plenty scary in the mountains when you are being knocked around in a rain squall and have to navigate strictly by sight. I've had a few scary ordeals thus far. One of my two helicopters crashed into a mountain while I was in Saigon. No one was hurt, and we salvaged some parts and blew up the rest so the enemy won't get it. You can imagine how delighted I was to hear the news when I returned! They already replaced the lost helicopter *(an obsolete H-21 anyway).*

The monsoon also makes working at this war miserable. You have to keep going, and it doesn't take long before everything is loused up, wet, and filthy. To add to the joys a monsoon provides, the snakes are starting to crawl in here to find warm, dry spots, and you have to flat keep your eyes open. There were 5 spotted and 4 killed this week alone. The one "Terrible Tony" and I saw was a type of viper that was more scared than us and fled before we had time to get scared. Two were baby boa constrictors. One, over six feet long, was killed in a ditch about twenty feet away from my quarters. A boa that small can't do much more than scare a man. But we also have a herd of huge rats gallivanting about the place, which the snakes eat, so I guess we'll have to work at getting rid of both snakes and rats. All you do is keep alert and keep proper precautions and you can avoid critter collisions. One good thing about the monsoon is that it cools us off, but the high humidity keeps everything damp.

Montagnard crossbows, a local weapon found in the tribes around here, have a particular appeal to Americans. The natives must think we are nuts as they watch us coming down the roads and jungle trails. There we are armed to the teeth with machetes, pistols, tommy guns, and grenades, and each of us was clutching a primitive old crossbow we just acquired. They make a good-looking wall-type decoration. All the flyboys and rear-echelon people who touch in here go nuts over the things and will pay up to three times what we bought them for. I'm told by a friend that he shot a bamboo arrow 3 inches into a tree with his at 50 yards. Pretty deadly weapon, though primitive. Europeans used them in the days of the castles and knights. The Montagnards use them for hunting, protection, and even against the Viet Cong. The Viet Cong even uses them when rifles are in short supply. How's that for warfare in the nuclear age? We buy all the crossbows around here and we sell them at 300% profit and then turn the profit over to Doc Smith so she can buy medicine to treat the Montagnards with. We proudly call her our own local "Tom Dooley." I might add she's our only available doctor too, and we've sent a lot of guys to her already since I've been here.

At any rate, after we cornered a few mystified Montagnards and bought their crossbows from them after much admiration, word spread across the mountain villages that the crazy Americans were going nuts for crossbows. This started a local industry. Before, each Montagnard fashioned his own crossbow for his own use. When word got around that we were willing to pay for those things, these guys started coming down out of the hills with three or four crossbows strapped on their backs, hunting for my fort. Once again, American business has spurred on the enterprising local industry and given the economy a shot in the arm! I think they think we are suddenly discovering a great new weapon and we are finally getting smart and stocking up on the ultimate weapon. Anyway, we've got a crossbow boom going on the side here, and under the direction of "Terrible Tony," these poor people don't know it but Doc Smith is able to buy more medicine to treat them with. It's not as much as we'd like to be doing, but it helps some anyway, and our job is to help fight the war.

We take a chloroquine tablet once a week as a malaria suppressant. No one has malaria yet, so the stuff must be OK. *(They said you still get malaria, but the pill, a big red rascal the size of a button, suppresses the symptoms.)* They also do a great job of binding up your bowels and causing constipation. But all you have to do is catch a mild dysentery, which is easy around here, and you're back to normal again. Quite a place!

Well, it's time to go over and shake out the pad to see what falls, jumps, or crawls out of it, and then catch some shut-eye. Being a craven coward, I generally sleep with both pistols, 100 rounds of ammunition, 2 grenades, a club, and a DDT bomb on the night table. The braver ones . . . come to think of it, there are not any—we all sleep that way.

Kontum, 11 June 1962

There is an argument about paratroopers versus Special Forces.[8] The fact is, they both do much the same thing in similar ways, and you can't say which is tougher. American and Vietnamese paratroopers over here are performing head and shoulders over Special Forces and the ranger units they are supposed to be training. I have a small group of Special Forces that operates in and out of here and a larger one at my Tan Canh outpost. These particular people don't show much and are more trouble than they are worth. They are making a bad name for themselves and us by their rowdy behavior. In guerilla warfare, you do that and you turn people to the Viet Cong, and these clowns are supposed to be experts! The other night, it was reported to me that a couple of them forced their way into a Vietnamese hotel in Kontum and started a brawl with the owner and some customers who didn't even try to resist. A crowd gathered and Viet MPs [military police] had to come out and get us to bring them in. Were we ripped at them! Good thing for them that "Terrible Tony" isn't here, and they moved out the next day. They got away before I could get my mitts on them, and I was mad!

Not long ago some of them were training a Vietnamese ranger battalion out on a mountainside in Cong territory. They were using a special type of rifle, the Armalite *(later the M-16)*, which was to be kept from the VC at all costs. *(They got the new small arms, while we used up the junk. Was it because they worked for the CIA and we were army? I believe so.)* At night 20% of the unit was to be allowed out on pass to a little village down the mountain. 80% were to stand guard. One of our captains and a Viet major made their way up to them one night for a look around. What a shock! The rifles were left scattered all over the place. Almost the whole unit had gone to town, including all of our Special Forces superheroes. Don Wood, the captain, fired a magazine into the air from his submachine gun just to shake them up, and the few so-called rangers that were around fled, leaving all the special rifles.

That's the Special Forces reputation over here at Kontum right now. They are arrogant, rowdy types, more show than soldier, and they are

---

8. *Webster's New World Dictionary of the Vietnam War*, 375.

getting people angry at us. Special Forces people, who had a reputation based on their press clippings, and who we all respected, as you know, are now resented quite a bit by the other guys who bear the brunt of this war without fancy green beanies and press clippings. And furthermore, the Vietnamese ranger battalions that these people are training are turning out to be poorly trained and poor combat outfits. One shabby Viet Cong platoon, poorly equipped, held off an entire ranger battalion with the finest equipment and Special Forces advisors for over 5 days! The Cong rabble was outnumbered 10 to 1! This ruined an operation where 3rd Battalion Vietnamese 40th infantry was tearing up a 1,000-man main force of the VC under tough tactical conditions. 3rd Battalion was outnumbered by about 100 VCs and still made progress in the attack. The so-called supermen loused up the whole show, and the Viet Cong escaped into Laos.

The comparison of Green Beret to paratrooper is not really valid. The Green Beret is a paratrooper or airborne ranger with skillsets for guerilla warfare missions. The paratrooper is among the world's best light infantryman of the twentieth century. Paratroopers fight in regular outfits in the main battle as parachute, air assault, or regular infantry. They don't require all the skillsets of the Green Berets, but the motivation and the physical conditioning are similar. The individual soldier skills and the group unit proficiency of American paratroopers cause them to be regarded as our elite. Seeing this bunch of Green Beret bad apples was a sobering experience. I met many later that lived up to the better reputation.

The Special Forces worked for the CIA, along with a whole group of aircraft outfits, helicopter and fixed-wing. This was a prime example of violation of the Principle of War called Unity of Command. We younger officers complained about all these sorts of things, and eventually a number of us resigned our commissions, as the ultimate appeal of policy from the officer corps. But the senior officers of the army did not appeal policy to the same degree that we did, so our sacrifice was a waste, except that by leaving the service, a number of us probably saved our own lives and our families' loss by missing a second, third, and even fourth tour of duty in Vietnam.

We have fans and we have showers. They ain't like home, but technically that's what they are, and they often work too. We have trouble with the showers, though. The absorbent walls are beginning to smell. I found out why today when I tailed one of my Vietnamese soldiers into the showers and saw him urinating in the shower. They think the showers are a toilet!!! What a scrub-down that inspired!

Yesterday I had to take two trucks and two jeeps down Route 14 to Pleiku to pick up the supplies they couldn't fly in here. When word got around that there was also mail on the plane, I had to turn down Yank volunteers. It was a very interesting ride. Route 14, though mostly tar, is one of poorest, bumpiest roads I've been on. We couldn't pour on the coal and rush on because of the road condition and the trucks with us. We had to slow down to a crawl in some pretty suspicious-looking places, so we had our thrills. I thought over and over of the many times the French got caught in the same places, but once again, I made the ride as a mere tourist. The country was beautiful and interesting and hid no enemies who were willing to cause any trouble, at least.

Every now and then we would roar by one of those strategic hamlets you've been reading about. This whole show reminds me of the Old West. The Cong are the Indians, the ARVN and us the army, and the people the settlers. Only three things are different. Vehicles instead of horses, jungle-type country instead of the West (more like colonial times), and a few newer types of weapons to go along with the primitive ones still in use. I know now just exactly how the soldiers on the old American frontier must have felt. Our situation is almost identical. Only instead of contending with a desert once in a while, we have to contend with the jungle all the time. *(Desert Storm vs. Vietnam terrain showed the difference in our abilities to find em, fix em, fight em, and finish em. Further, our foe in Vietnam, to their credit, were to date the only foe of our illustrious military history who finally won the war against us.)*

Resupply operations to the advisory teams in the field were real military operations and plenty hazardous. One of Vietnam's French-built highways, Route Nationale 14, went right past our base camp on into the northern wild country. For a number of miles, the road was a two-lane blacktop, not unlike a country trunk-type road back home. Then for some miles we had a mud, dirt, or dust road depending on the season. Finally, it became a jungle trail.

In *Street Without Joy* (1961), Bernard Fall wrote about the ordeal of Group Mobile 100 in the French war eight years earlier along this road. Khon Brai, one of my northern-most outposts, was where it started. It ended on Route 19, about 50 miles south of us. Skeletons of the French

American–built equipment were still visible up and down the route. It was an epic fight between Viet Minh and French heroes. The local Montagnard tribespeople never forgot the French. They carved wooden totems in the likeness of French soldiers, which we would be astonished to come upon as we entered their villages. They saw us as replacements for the French, to whom they were loyal. As an amateur historian, I could discern more to this Montagnard thing than was then known by modern scholars, and I truly wondered about their connection with our Native Americans and some long odyssey across Euroasia, etc., etc., etc.

A typical resupply operation by chopper would mean that one ship would carry the supplies in and bring out any casualties, while a second gunship would fly cover overhead. Probably my first resupply mission as the Kontum base camp commander was to resupply the advisors at Khon Brai. We flew up there in two "flying bananas," and the ship I was in flew "shotgun." I had a 16mm camera and filmed the other ship going in and doing the resupply mission.

Discipline was a big key, and it worked both ways. You could be too careless, or on the other hand you could get too jumpy and shoot up the wrong target. Thank God for good training and discipline. The war was hotting up, and I was taking some supplies north one afternoon peering intently into the brush with my burp *(submachine)* gun at the ready. The bushes suddenly swayed back and forth as we approached, and I instantly prepared to fire. Out of the bushes came a young Montagnard mother with her baby papoosed on her back, as I had seen in pictures of Native Americans. To this day I am haunted by the horrible thought of what if I wasn't so well trained and disciplined? How could I have lived with that if I became too jumpy and opened fire? How many young American soldiers after me who had not the training, discipline, and experience I have made that mistake? My God, we were there to protect them; in the mistakes of war, how many did we kill?

There is nothing you could really call a rear echelon—no secure rear areas, very few fortified areas which guarantee absolute safety. When you go on a supply run in the war, your chances for a good fight are better than when you go out with a rifle company *(circa 1962)*. A hint of suspense covers the whole area. After a while, the French had enough of it. We at this time are in better spirits. *(We were just starting.)* But we are not battle-worn over here yet, like the French were. And we know we have the mighty resources of our own country, which can come to our assistance at any time. *(That was the attitude of a typical American captain in middle management at that time. A few weeks later, notice the change. Decades later, a careful reading of the now declassified official documents of the time will show more of the story. How much more have they hidden away somewhere?)*

No one will ever be able to compare this war with World War II or Korea. It's weird, it's different; combat in our sector has not seriously touched us as yet. When it comes, it's short, violent, and infrequent. At this time, it's almost entirely Viet against Viet Cong, and only once in a while does an American get shot at. *(But see reports below from a few weeks later.)* That usually is on a road or in a helicopter.

Action up here looked like it might be stepping up a few days ago, with sharp fights to the north and south of us. Near Ban Me Thuot on Plateau du Darlac a big fight started that's still going on. Another fight started by Khon Brai on Plateau Jea. Ambulances roll by here a couple of times a day with the wounded. This is a fairly good sign. If the Viets can get single ambulances down the road *(Route 14)*, they must be in control. Reports are that the Cong is losing 3 men for every Viet. The annual Cong drive for Khon Brai appears under way. Last year they took it and swept on to the outskirts of Kontum. My place was about one quarter its present size then. They ignored MAAG and hit a Civil Guard post visible across the field several hundred yards from here *(later one of my outfits)*. Some stuff flew around here, stray rounds, ricochets, etc., but the Yanks around here were out of it. The Cong failed and scattered back into the woods. This year everything in their way is bigger and tougher. *(But so were they. The year-by-year story.)*

Khon Brai, while small, is vastly improved. Regulars man the works, and friendly patrols roam the surrounding hills. Tan Canh is heavily manned by Viet regulars of the 40th, with 20 of my guys *(Yank advisors)* all heavily armed. Dak To and Dak Ro Tah are stronger. The Civil Guard *(my next outfit)* posted across the field from me has better-trained personnel. I have a large company of Americans and Viet regulars of the 40th in my fort, and the Kontum area is crawling with Civil Guard and units of the ARVN 22nd Division. *(Later I would take over advising the Civil Guard and SDC [Self-Defense Corps].)* The enemy may take some of the places north of here, but what a fight it would be. Kontum, I think, is just too much for the enemy, though they may harass us some.

Most of the fighting, though, continues in the southern Delta regions. My buddies down there are already veterans of many fights. Up here when we go out, we expect it, but it rarely comes. We get the eerie feeling of being watched, but that's it. I'm convinced the Cong is trying to spook us up here, while the main effort is south. It's easy to see how they can succeed. After a convoy ride through the jungle you virtually have to pry your hands from you burp gun. You scan the ground at treetop level from a helicopter and you just never know.

Every so often someone gets shot at. They have enough strength around here for more than that, but it looks to me that they do not understand Americans. They are trying to get us to look jittery in front of the people. But it ain't happening that way. We scare and feel pretty eerie, but we have a peculiar American way which fools foreigners. Our guys make small talk, kid around and joke when the going gets scary. We look like we are brightening up instead of tensing up. Or we look like we are back home riding along in the convertible. We wave and grin at the surprised natives. We look totally self-confident to the Viets and the Montagnards. Viet and Montagnard soldiers like to go out with us. I'm pretty sure of my theory for this reason. *We* know when we are scared. But *they* don't. This is pretty important to us.

This particular Cong effort, a very subtle one, to discredit us, is flopping. That leaves them with the nasty alternative of trying to defeat us in a fight. They have avoided us *(the Yanks)*, and they have avoided ARVN to a large degree because (at the moment) they fear a fight with either us or the ARVN. They'll fight the Civil Guard, and they love to fight the Self-Defense Corps. *(I later advised these guys, and they weren't so easy either at the time.)* But when the Cong takes us on, they tackle well-armed regulars. What's worse for them, they might get America mad enough such as they have never seen before. They know it. They seem content to play spook with us.

That's fine with us. We figure if we can get ARVN to get out and win this war, that's what we came for. *(Notice how our attitude changed as we realized we were left out to dry by a conniving American government and a passive, ignorant public.)*

Kontum, 17 June 1962

General Decker (chief of staff of the army), General Harkins (chief MACV [Military Assistance and Command, Vietnam]), General Timmes (chief MAAG [Military Assistance Advisory Group, Vietnam]), and various other big-wheel types spent the day here Friday. In addition we had Lt. Col. Lucius B. Clay, Jr., son of the famous General Lucius B. Clay, who figured in the Berlin crisis. My fort was crawling with dignitaries Friday. Col. Clay stayed for a few days and read my books and notes on guerilla warfare. I sold him some crossbows for Doc Smith. He was a good guy. He was shot up, not seriously but painfully, a few weeks ago on two different occasions in a helicopter. The other VIPs stayed half a day, ate lunch, and departed. Generals Harkins and Timmes were friendly, but Decker was a sourpuss. He's one of those types that doesn't even look at you unless you're a colonel, I guess.

"Terrible Tony" came back from the Philippines yesterday in a good mood. No violent outbursts as of 19:45 tonight. He says I'm doing a good job. But he can tell you that one minute and fly into a tantrum the next. What a guy!

We had a jolt the other day. Received a message from Saigon that a sergeant was killed and a captain and sergeant wounded in action 360 miles northwest of Saigon, near the Laotian border. We checked and rechecked the battle maps and plotted the location around Kontum. We couldn't figure out how Saigon found out before we did. Worse yet, we were sweating out our guys who were out that day. We had three similar teams in the vicinity. It was also reported that 6,000 VC in the largest concentration ever in this area *(to mid-1962)* were gathering to slam us a good one. All of this information coming in caused mounting concern. I ordered the grass and brush around our fortifications chopped down with machetes to a distance of 300 meters. This makes it difficult for anyone to get close unobserved. We were much relieved when this bad news was watered down for Kontum at least. All our guys came back. Apparently the group that got clobbered was from I Corps, operating in our neighborhood. Estimates of the Cong force gathering to the north have been revised down to the point where we aren't sure but we think we are talking about hundreds, not thousands. Everything remains well in hand.

This place I'm at here, the "Kontum Hilton," is located on the northern outskirts of Kontum. Some of us run the "fort" itself. I command the whole thing, and these people work for me. Most of the people who stay here are teams of advisors who go out and advise ARVN, Civil Guard, and Self-Defense Corps units. Some of these teams stay here all the time; some spend most of their time in the field using this as a home base. The Old Man, "Terrible Tony," runs the whole show from A to Z, whereas I run the fort. I have about 20 Americans, including two other captains and two lieutenants. I also have a 50-man platoon of infantry (ARVN) to man the defenses. I also have 15 additional ARVN soldiers helping with the housekeeping chores. Then there is a 20-man platoon who perform construction labor tasks. Finally we have about 30 Vietnamese civilians employed here. Altogether we have about 14 buildings, 9 large tents, 4 water towers, 3 mess halls, internal electrical generators, and a good-sized signal setup. I even have a small PX [post exchange] and 2 bars and a small pool hall, barber shop, and laundry. I also have a motor pool that services about 30 vehicles. And we usually have 2 to 4 helicopters to take care of. In addition to this I have a smaller edition of this place with facilities for 20 men at Tan Canh, located just outside Dak To. It's a rough, busy job. I've always been a hustler, but "Terrible Tony" still stays a step ahead of me.

This place is surrounded by a barbed wire fence entanglement with bamboo spikes outside of it. We have trenches, sandbag earthworks, a watchtower made of concrete, log-supported earthworks, and six well-sited machine gun emplacements. We have outposts out on the barbed wire every night, with roving patrols at irregular hours. We have ARVN artillery concentrations plotted all around us. Every man has a specific place to go in case of attack. We have a large number of Thompson submachine guns, a few grease guns, many carbines, several BARs [Browning automatic rifles], and many M1 rifles (*all old, obsolete weapons*). I have a command post and reserve position in the center of the compound, where a counterattack force with 3 machine guns will assemble. Everyone sleeps usually with a pistol, two grenades, plenty of ammo, and one other weapon, usually a submachine gun. Plus that, some of us carry a club to bed for unfriendly critters that might get in during the night. There are certain taboos around here for safety's sake. There's not much roaming around outside the barbed wire. Anyone stupid enough to go running through the place at night (as some young soldiers found out to their great fright) is likely to find himself a dead man. Civilians leave at night.

Basically my job is to keep all this functioning as a base of operations for the advisors who operate out of here. I have very little to work with, and it's rough and usually thankless. When things go like clockwork, that's expected; when the least little thing goes wrong, even if it's a matter of personal whim, the world comes crashing down.

I need all the Vietnamese I can speak to do the job, as you can see from the number of Viets who work for me. I'm actually one of the few Americans in Vietnam who commands Vietnamese troops. I'm also the only American in Kontum Province who speaks Viet. Because of this fact alone the Viets around here give me anything I ask for, and that's a great advantage. The vice governor of the province, Ong Taum, who is actually the top Viet in this area (the governor is merely a political figurehead), has turned out to be a great friend. He says, "For you, who speak Vietnamese better than all Americans, I do anything. You are a Vietnamese poet!!!" Just what the hell he means by that would be hard to say as an American. But Vietnamese to Vietnamese, I think it's a compliment. *(In a few months' time I came to appreciate the culture of Vietnam.)*

Anyway, the other day I asked Major Murphy to get some road-building equipment for me because I have to build a road, among other things. He got flat nowhere [with the suppliers] until he mentioned who the stuff was for *(meaning me)*. The then vice governor pulled the equipment away from an important airfield project and delivered it straightaway. The airfield was immensely more important, so I had to ask him to please not be so hasty about fulfilling my request. I would have been perfectly happy to get the equipment *after* the airfield was improved.

I also discovered that my Viets are scared to death to be chewed out by a Westerner, and when a Westerner chews them out *in Viet*, they really cringe *and do not hold the usual grudge.* I can generally count on 100% loyalty and no desertions to the VC, even after raising terrible hell and driving them far harder at work than their own leaders move them. This is apparently because I simply speak the language and show a little interest in them. It is easy to see from what they seem to expect from me how the French treated them. The ability to speak Viet, and I'm improving a little each day, is a great tool in accomplishing the mission. I catch myself more and more often in earnest discussions with Viets while other Americans around stand there helpless. We could accomplish a lot more if we could all speak it. We have plenty of "ugly Americans" around Kontum alone because many can't speak the language and could care less. *(The early groups of Americans probably had more empathy for the Vietnamese culture and people than the Americans who came in later. And we early ones probably got along much better. Part of it, of course, was the newness of it. But even in those days we had a number of arrogant racists who must have hurt our cause.)*

Individual and small-unit punishment meted out by Vietnamese officers and non-commissioned officers was a shock for me at first. I saw a unit lined up in the heat at attention being belabored by an angry leader who didn't hesitate to beat the men with a stick while they stood at attention in formation. One technique for individual punishment was a hole in the ground under the hot sun, covered by barbed wire. The culprit stayed in there for a specified period of time and had to make waste in there with himself. There is a real culture difference in these kinds of situations.

Kontum, 24 June 1962

I have to go up to Tan Canh by armed convoy this afternoon, which is always a stimulating experience. We have some new grease guns in, so I'll try one of those. *(I became pretty good with the grease gun, which fired .45 caliber ammunition, same as the army .45 pistol. I could squeeze off single rounds and hit a bullseye at one hundred meters.)*

This is a very strange war. One night last week the Viet Cong moved into Kontum and occupied our airfield and the bridge over the river on the road to Pleiku (Route 14). Not a shot was fired. Come morning, a company of ARVN paratroopers jumped in and took it all back, all within sight of my place here. Still not a shot was fired. Very strange indeed! One feels a little weak over here because of the climate or the pace or both. We generally seem to attack the day with more vigor than the Viets do, and I think they find us hard people to work with.

The colonel came back from Tua Hoa [Tuy Hòa] today. He wants me to rush to Saigon to check on some things tomorrow.

Saigon, 27 June 1962

I have been catching up on my rest here in Saigon in an air-conditioned hotel room, so I feel as though I have been living high on the hog. I feel as though I have regained my strength and am ready for another go at it again.

I see that the war in Laos has ended with a rather shaky arrangement.[9] The Communists, now in the coalition government, have just the setup they want for complete takeover, I think. We will find out soon enough up in Kontum when we see whether or not the Viet Cong continues to move men and supplies in from Laos up our way. My guess is that they will. I think our government succeeded in getting a breathing spell for a short while only to let us in for more trouble later.[10]

Saigon weather, even though fairly comfortable now, is always warm, muggy, and humid. There are always a few showers floating around. The city has many odors and seems very dirty. The people vary in their friendliness toward us, and many seem outright hostile. There is a definite air of tension about, and one of suspicion as well. I do not care to be wandering about the streets here, and I do not especially like the people in Saigon itself. I like the people up around Kontum a great deal better. Saigon has a good many charming buildings and fine tree-lined streets of French influence. But even this does not overcome its unpleasant aspects. It is difficult to make friends here for the sake of having friends. The people here only deal with you for business or because they want something. They are very suspicious, and I don't like that at all. This is only in Saigon, and for that reason it ranks at the bottom of the list of the cities I've been in. It's full of intrigue and filth, neither of which is necessary if the people would put a stop to it.

The government *(the Diem regime)* is a police state which delights in harassing Americans to remind us who is boss. This keeps the pro-government people away from us. The Viet Cong elements also avoid us and throw potential danger into the picture. Others are only after the Yankee dollar. If there is anything that makes me mad it's to see a local go by and look at me with hate in his eyes. The stupid slob hates for the sake of hating and will live and die with hate in his heart without taking the trouble to know what he's hating. I think many of these people hate us for the love of blaming everything on somebody else. Hating Americans is fashionable these days. I don't think we can prevent this type from hating us. He wants to hate somebody. If he has nothing, he gloats in his belief that the Americans are getting rich off of him. If we give

---

9. See H. R. McMaster, *Dereliction of Duty: Johnson, McNamara, the Joint Chiefs of Staff, and the Lies That Led to Vietnam* (1997).

10. See chapter one, "Straight to the Brink Over Laos," in John M. Newman, *JFK and Vietnam: Deception, Intrigue, and the Struggle for Power* (1992).

him something, we hurt his pride and he resents being helped. The type of understanding this guy needs is one which even we Americans in our paradise have not fully grasped. The feeling of hostility, suspicion, and isolation which this gives us makes it that much more difficult for us to try to help these people. *(One wonders how this kind of person felt about things after a generation of Communism.)*

Kontum, 3 July 1962

I've had to work about 17 hours a day since I got back from Saigon and it's rough. Things aren't going too well in our mess. My predecessor left me with a can of worms, and things didn't get any better during my first couple of weeks when I was learning.

Kontum, 6 July 1962

Well, the past week has been a corker! We developed some problems in the mess portion of our operation and also with our electric power sources. We have been jumping! It may take us some time to get these problems straightened out, but we appear to be on the road to recovery.

In addition, one of our officers got ambushed out on the road to Tan Canh and we had a jeep messed up a little as a result. He ran over a hand-operated mine that a Viet Cong soldier set off a second too late. The explosion ripped up the back end of the jeep, and then the shooting started. The major and the men with him returned fire so violently that the Cong broke it off and ran away. It didn't take long before the whole ARVN army was chasing through the woods looking for those VC, but they couldn't find them. The only ill effects were the major's map case, which had a neat bullet hole through it. I went up the road at treetop level in a helicopter a few hours later and we found nothing and saw nothing suspicious.

Saigon, 7 July 1962

I seem to be making a second home down here. There is much pressure in this job I have now, and I don't like the job, so the pressure adds to the unpleasantness. My biggest headache is running the mess portion of the operation. This is a strange setup which resembles running a club in the middle of a jungle war. Everybody pays in so much a month and we run our own mess. The thing is big enough to be classed an $85,000–$90,000 business, so it's no small operation. *(That's what I* *thought at the time, but now I guess it was probably about half a million in late-90s dollars—a small business. [In 2025, the equivalent might be close to $1 million.])* I started out with a minus net worth of $2,300 and too much inventory.

Have just returned from an interesting evening walk. Many types of shops about. The French people here seem very much in evidence for a change. Major Rovegno is down with me. This afternoon we took advantage of a good opportunity to relax and we went over to the Cercle Sportif to sit beside the swimming pool, have a beer, and look at the French girls.

Saigon, 9 July 1962

This time I am staying at the Hotel Majestic, one of the nicer French hotels. This room is a little crowded with 3 of us, but it is air conditioned so that it is almost too cold for us to stand. The place has a nice French atmosphere, with French people all around. It overlooks the Saigon River waterfront. The dining room up on the 6th floor overlooks the entire scene. Across the river, it is hard to believe, is VC country.

Another big load of guys came in today, including another OCS [Officer Candidate School] classmate of mine, Ernie Cherry. It is difficult to realize that I look at these guys with only ten months to go as sort of old-timers. They are asking us all the same questions we were asking.

Like I figured and may have mentioned a few weeks ago, the Communists were fighting in Laos several weeks ago and are now coming into Vietnam in the Kontum area. They estimated 8,000 of them. We don't know what they have in mind yet. They may just move through Kontum Province to operate elsewhere, or we may be in for it. No sweat.

I think it was in July 1962 when a number of aircraft landed at Kontum bringing our advisors from Laos who'd had to leave with the neutralization of Laos.[11] These guys first went to Saigon and then were reassigned in Vietnam.[12] Until that moment, it looked like the Kennedy administration was developing a uniform policy for Southeast Asia, with Laos and South Vietnam holding the line, backed up by the US and SEATO, and with Cambodia neutralized on a friendly basis. The way the line was, we had

---

11. McMaster, *Dereliction of Duty.*

12. These must have been Bull Simon's White Star teams. See Jane Hamilton-Merritt, *Tragic Mountains: The Hmong, the Americans, and the Secret Wars for Laos, 1942–1992* (1992), 119.

North Vietnam with the longer hostile border, a good chance for containment! Any child or an idiot could see that a neutralized Laos was a Communist playground—the ten thousand Pathet Lao suddenly becoming ten thousand Viet Cong, and the entire western border of Vietnam given to the enemy for the Ho Chi Minh Trail. (Exactly what immediately happened!) It was frankly terrifying to suddenly realize that we had been so betrayed by our own government at the time, as had the South Vietnamese.

Saigon, 9 July 1962

Things up our way are getting livelier and that's OK with me so long as they keep missing us. Some of the Reds who were fighting in Laos a few weeks ago are now coming into Kontum Province, since the Laotian thing is over with. That gives us 8,000 more of them to fool with. I knew this was going to happen weeks ago. The question is whether they are going to pass through Kontum to operate elsewhere or whether they are going to heat up our already warm sector. So far we have been lucky. The guy I came down here with got ambushed last week, and like I've been saying, the enemy did a poor job. They tried to blow a mine under his jeep, but it went off behind him, did a lot of superficial damage, put a nice hole in the road, and didn't hurt anyone. Then they shot up a storm and kept missing. The major and his compatriots put up such a violent firefight that the Cong fled. A large relief column combed the surrounding area, but it was the usual story: they got away. I had to travel the same road a couple of hours later and nothing happened. This is everyday hogwash by now and doesn't bother anybody too much. We compare ourselves to you turnpike drivers back home and consider ourselves safer. *(Well, wasn't that a piece of reassuring propaganda?)* Personally, I much prefer a drive up Route 66 to a jaunt on Route Number 14 to Dak To or Pleiku, but c'est la guerre.

One evening I was returning to my room at the hotel and noticed a foreign officer coming down the other way and stopping at the room next to mine. He was a Polish captain. He saw my name tag, FRANKWICZ, and we both paused a moment, sizing each other up, the third-generation American captain of Polish descent and the Communist Polish captain, cousins, so to speak, made enemies by history. Then not speaking, we each entered our rooms. He eventually to return to Communist Poland and me to return to the good old USA.

Saigon, 10 July 1962

I seem to be spending most of my tour in Saigon! We couldn't fly out today because of the weather around Kontum and Pleiku.

Part of my duties as detachment commander has to do with running messes and bars. I refer to myself at times as a combat club officer. I have finally come to regard the thing with a sense of humor most of the time, but it has been a living nightmare for a while. This is an $85,000–$90,000 business *(about half a million 90s dollars)*. Everything requires much more planning, and you have to cope with many kinds of emergencies. *(Like losing a planeload of supplies that belongs to us personally.)* I generally run the whole thing; I keep the books, handle all the money transactions, and am responsible for it all. I have several people to assist me.

I am surprising myself at my ability to handle this thing thus far. The guy who had it before me turned it over to me on 10 June under a series of wrong assumptions. His financial statement for May showed a net worth of $4,500, liabilities of less than $100, and he advised that we buy heavy on food inventory to stock up for the monsoon season. There were other things too, but the guy was wrong. His net worth was about $500, or $4,000 less than he said, he operated at a loss in May, his liabilities in fact were over $4,000, and several other things. It took me the remainder of June to figure out what was the true picture and at the same time set up some adequate controls, of which there were none. I was stuck in June having to operate on the assumptions this guy and the Council gave me to work with, and I very nearly got us into some serious trouble, but I caught it before it got too bad. I was still finding liabilities for May in July. What was worse, under the operating system and assumptions I had to work with, I was incurring even more liabilities blindly in June.

By the end of June I was able to construct a true financial statement of how we stood, and we were quite shocked. We found ourselves with liabilities of over $10,000 and a minus net worth of $2,300, with an operating loss of $3,000 for June! Was the Old Man [Tencza] ripped! It didn't take me long to make matters worse on the basis of incorrect records and the lack of controls I had to start with. It was too bad it took me a couple of weeks to work it out. Everyone had so much confidence in my predecessor that I was more or less high-pressured into going full steam ahead in his footprints, even though I warned them we had better hold off for a month or so to see where we were.

Well, come the first of this month I installed all my controls, paid off $4,100 worth of the liabilities, discovered we could operate on our inventory for a month without incurring much more liabilities, and next month we would raise prices to cope with remaining debts. Also, everyone now knows where we stand.

To add to our woes, it looks as if we lost some inventory along the way, and it can't be pinned down. The Old Man was so mad he threatened to relieve me, and wouldn't have cared much if he did. He threatened to send me, my predecessor, and all my people straight to Hanoi, and he roared that we would regret the day we ever met him. That day, I regretted the day I met the old war horse. But he [eventually] calmed down and realized what happened, and now we are settling down to dig out of this thing.

After all, he was fooled too until I told him about it. He's mad at me because I was at the controls of a ship without a compass and didn't help matters any, and in fact made the situation worse until I was able to get a compass on the ship, so to speak. I really don't know how justified he is in being ripped at me for a hand in it, other than the fact that I theoretically had a hand in it, and I was the guy that brought the bad news. If things really get down to the worst, which I doubt because the Old Man can be pretty fair in spite of himself, I would be really happy to see someone else take over. But fortunately and unhappily at the same time, as one of the guys said, I'm probably the only one who can pick this thing back up again. I feel as though I comprehend what has happened, what must be done, and what I am doing. If my program succeeds, now that it has a chance to take hold, I will have this thing recovered in a couple of months, God willing. If I am successful at running this thing and also performing my duties as detachment commander, I should be able to do most anything when I get out of here. The army's methods of doing business are too inflexible, and the army doesn't allow much for mistakes, and in fact, the army often doesn't know *who* made the mistakes. *(Actually, we were being supplied by the navy's system, and we were running an operation we had never seen before, and doing it in combat. Once again we found ourselves in another inconceivable situation, in which normal army mess operations were not instituted and instead they were making combat officers jump through hoops.)*

When we went overseas to Germany in 1956 as newly minted 2nd lieutenants, they sent all of us paratroopers to the [10th?] Division, and the ROTC

*non-airborne-qualified guys to the 11th Airborne Division. So the para-
troopers received officers who didn't jump, and before long rumors were flying
around about the discipline problems in the 11th Airborne. After a year on
the line I left the [7th?] Infantry for Headquarters, Seventh Army, in Stuttgart, and among other duties, they made a bar and mess officer out of me.
I had no training in business management or accounting whatsoever, but
excellent training as an infantry officer, which was where I wanted to be and
where I did well. The army had scores of ROTC business school graduates
nearby who would have preferred the job to freezing out in the winter with the
infantry. I did not do well at the job, and later, in St. Louis, I took accounting
and management classes at the university at night to better prepare myself. So
when they stuck me with this job in Kontum, I was able to do it.*

11 July 1962

As fate would have it, our plane was delayed by weather yesterday and I
ran straight into Milt Craddock, who is up from Baria [Bà Rịa]. Another
old buddy, Jim Presley, is down there with Milt doing the same thing I'm
doing. Milt has a rough go of it down where he is in III Corps area. He
advises a battalion of the 48th Infantry and has been in frequent combat.
He has had many narrow escapes and hopes to get a little safer activities
soon by taking over Presley's job when Jim goes home in a month and a
half. When I told Milt about my job he most emphatically stated, "Don't
try to get another job! At least you have relative safety where you are, if
nothing else." He's right.

A couple months of combat have put Milt on edge a little. He has
some interesting stories to tell. He likes working with ARVN troops and
feels a great respect for them. He says one outfit he was with would rank
with our best. *(He was an NCO in the 82nd Airborne and a company officer in the [7th and 10th?] Infantries with me.)* He also thinks South Vietnam will win out in the end. Milt says the presence of Americans is not
too valuable from the standpoint of helping operations. Sometimes our
guys just get in the way. But the mere presence of Americans watching
has made these proud Vietnamese soldiers go all out to look good. Careful tact in dealing with them, and a well-chosen word of praise here and
there, is really helping these Viet leaders out. What seems to get to him
the most is that he often gets a trapped feeling, where he wants so much
to go home, but there he is right in the middle of it, not knowing how or
when or if he's getting out. *(I served much of my military career with Milt
Craddock, who was a soldier I much respected and a good friend. I met him*

our first day at OCS at Ft. Benning in 1956. He and another SFC (Sergeant First Class) from the 82nd Airborne were sitting next to each other on a bed looking plenty worried. I'd just come in from being an instructor at advanced infantry training at Ft. Dix, and when I saw these two veteran paratroopers sitting there like that I wondered if I shouldn't really get scared. We went through OCS together, along with some other guys mentioned here. After OCS, some of us went through paratrooper training and jumpmaster school, and Milt and the other paratroopers welcomed us into that elite fraternity. Then we went to Germany and we served in Schweinfurt in '56–'57 and in Augsburg in '59 as company officers in infantry outfits. Again later, back from Vietnam, we were classmates at the Advanced Officers Career Course at Ft. Benning. Milt's hair had turned gray during his ordeal by fire. On one occasion, his unit was passing through a village and a small child threw a hand grenade at them. On another occasion, one of our classmates from the Presidio language course, a West Pointer, the one who'd got the airborne battalion, was trapped with his battalion in a rice patty, surrounded in a tough fight. Out of ammo, they prepared to fix bayonets and try to break out when Milt's outfit broke through to get them out.)

Kontum, 13 July 1962

The heat you are experiencing in St. Louis right now is probably worse that it is here. The humidity here is terrible, but the temperature does not get unbearable too often.

Nights I often have to pull up a blanket for a couple of hours. Saigon is even a little more comfortable right now. The monsoon is acting very strange this year, coming and going sporadically. We had one period of almost two weeks without rain. Then it will rain from several times a day to continuously for a few days and then strangely clear up somewhat again. The Vietnamese say it is a very strange year. It's messing up our supply runs, which I don't like, but more is getting through than might be. Mail is still fairly regular.

Kontum, 18 July 1962

Well, things are picking up in the war around me, as you may be finding out from the news by now. We suffered some casualties last weekend among our people on operations, which shook us all up deeply. It seemed like one thing after another for a while there, and I was beginning to wonder when all the bad fortune would stop.

We lost another helicopter shot down by enemy fire. In that action we lost our great commander, Lt. Col. "Terrible Tony" Tencza, who apparently climbed out of the wreckage and died fighting. Three others—CWO Goldberg and Specialists Guthrie and Lane—died with him, along with Vietnamese officers. Major Council and Captain Hoa managed to get out, wounded and battered, but alive. We have our story from survivors and from the friendly forces that came upon the scene much later. The colonel had been on a recon mission investigating large-scale Cong movement into our area from "peaceful," "neutral" Laos.

More details about what happened: In July 1962, an enemy main force battalion got into our area and occupied the attention of the local leaders. On Sunday morning, July 15, 1962, I was sacked out, and Lt. Col. Tencza and some others walked past my "hootch" (sleeping quarters) on the way to the choppers. The colonel said, "Let this guy sleep; he's been working too hard," referring to me, and off they went on a recon mission with two choppers, looking for the enemy battalion. Later in the morning, I suppose after church, we received a message in the GP [General Post, which handled telegrams] that they found the enemy and the colonel's ship was shot down. A bit later the other ship came back for fuel, and there was much consternation, as we had nine guys on the missing chopper, including the colonel. Nothing like this had happened before in the Vietnam War. Until that day we were taking casualties one at a time. Now, for the first time, it looked like nine in one action.

For some reason, Major Reilly, the 22nd ARVN Division G-3 advisor, came to me as the base camp commander and told me I would have to decide whether or not to commit and lead the US troops in the base camp as a relief force to recover the lost men. I think I remember the chopper company guys from Pleiku had flown up to Kontum and were in much anguish about their buddies in the lost ship and were ready to go. We had about one hundred advisor and base camp personnel that we could take as a strike force. Major Reilly and I pored over maps in the portico near the officers' mess while I agonized over the decision, which somehow seemed to be mine as the base camp commander and the senior survivor in the chain of command. An unseen power seemed to be whispering to me, "Don't go . . . do not go . . ." Finally, torn with emotion, but not able to show it, I said, "Major Reilly, we can't go. Our mission is to be advisor to these Vietnamese outfits. If I take these guys out there to try to rescue our guys and get us all

*killed off, our mission in this area will be ruined for the foreseeable future."
Major Reilly said, "OK, that's it then. We'll just have to wait for the
ARVN to work their way in there to them."*

*Sometime later, nine bodies were brought in, including Lt. Col. Tencza.
A day or two later a live survivor was found. He said he'd made his way out
of the ship into a nearby river and watched from the undergrowth while the
enemy pulled Lt. Col. Tencza over to a tree and then shot him. The enemy
in battalion strength (six hundred to eight hundred men) then just camped
for many hours, waiting to ambush us if we came. Had I chosen to bring
in the impromptu task force from the Kontum base camp in the Pleiku
choppers, I would have led us all into an ambush and been some kind of
Little Big Horn Custer-type of headline, the first of the war. I listened to
the "voice" and our losses stayed at nine. Thank God (or the voice). I believe
now that was the Spirit of God or one of His angels speaking to my spirit
and working intuitively within me. This was part of the beginning of our
personal war, and it was also a key part of the opening of the drama for all
Americans, though few took notice. That was on July 15, 1962, our first
wedding anniversary, and I was left with the memory highlighted each year
thereafter at that time.*

They managed to pull out one major alive. To add to the nightmare of
that week we lost one more American and 24 Vietnamese the next day.
Then on Tuesday two wounded American majors were brought in. I hope
we never go through this again; it was a nightmare.

We mounted counteroperations by flying in ARVN Ranger units, and
this brought our second disaster upon us before we regained our compo-
sure from the first one. A C-47 loaded with troops crashed on takeoff at
our airstrip, killing 21, including one American, and critically wounding
4. (*Later research in the mid-nineties suggests that this was a CIA operation and that the C-47 was part of Air America. There certainly were a
number of separate wars being waged, each with its own chain of command.
It's plainly criminal that such a situation was allowed to happen. We
needed one unified command with clear orders, and one tough warrior for a
commander.*)

Just about in the middle of this we learned we lost another captain in
an action to the south of us. Then reports came in that a B-26 bomber
and a C-123 cargo plane were down not far away. And as the nightmare
ended, two wounded Special Forces officers were brought to us.

Nothing hurt me more personally than the loss of "Terrible Tony." The Old Man caused me much grief, but we were close, and he was probably the roughest, fairest, best commander I've ever had. He taught me much in the short time I've known him. It happened on Sunday the 15th of July, our first wedding anniversary. As dearly as I love my precious wife, I couldn't even think of our first anniversary that day. We hoped against odds to bring the Old Man out OK, until at nightfall they brought in his body.

Operations go on, but not the same. I don't get harassed as much as before, but then no more do I see that familiar swagger and have him put his arm on my shoulder when he finishes roaring and says, "Don't worry about it, Mike, we're making progress." None of us ever will forget "Terrible Tony." He was a man and a soldier. He was a real leader. Not always right, but right enough of the time that the way he drove us he was head and shoulders above most. (He was one tough paratrooper.) As one of our guys said, "The Old Man died with his boots on, and that's the way I think he wanted it."

(I felt like Col. Tencza and I were growing a pretty close relationship. The army was like that. It was usually with peers, but it also happened with some bosses and subordinates. With some peers, I served with them together in units or in school several times in ten years. I knew Major Fisher when he was a company commander in Schweinfurt and then years later as a battalion commander at Ft. Leonard Wood. Col. White was a higher-ranking staff officer at Headquarters, Seventh Army, in Stuttgart and was my regimental commander at Ft. Leonard Wood. The good soldiers always had a lot of good stuff to rub off on you.)

This is war and we have to expect these things. There is no easy way to express oneself. [After Tencza,] Col. Read, a much more mild-mannered individual, took over. We are not entirely sure what to expect from him, but I think he will be less of a driver and fair. Col. Read is from St. Louis, and his family lives not far from Peggy *(my wife)*.

[Editor's note: The loss of Lt. Col. Tencza not only shook up the Kon Tum base but also ushered in a different phase of the war for the people stationed there.]

CHAPTER FOUR

THINGS GET HARDER

JULY 1962–JANUARY 1963

Kontum, 21 July 1962

I have built a road and a couple of buildings and started major renovations on others. I also built two helicopter landing pads and landscaped (yes) the whole area. The financial problems are on the mend.

25 July 1962

(A couple of days before I wrote this letter home I had received a garbled telex: "#@%&...mother and daughter doing fine...++#@$...7lbsl *&#," and that was the way I found out our first child, Michele, was born.)* All day I've had to say over to myself, "My own little daughter, my own little daughter!"

One of the chanciest elements is whether or not planes come in now. The food supply is down to only a few days, and we've had no mail or movies in days, and the liquor supply is getting low (ammo OK). After all the problems we've had this past month it will be especially tough to run out of food and have to scrape around for chow of inferior quality and possibly have to dig into expensive C rations.

The Air Force, as I recall, flew in two C-123 troop transports per week with our supplies. They ran these around from airfield to airfield resupplying the MAAG detachments around Vietnam. When the weather was bad, they just wouldn't come, and when we got desperate, the army came in with their Caribous [C-7s]. When the weather became still worse, only Air Vietnam would fly in their old DC-3s. I complained only once to my congressman, and it was about being "abandoned" out in the jungle with my troops. He forwarded my complaint to the Pentagon, and an Air Force colonel in the Pentagon sent a letter back through the congressman, refuting my complaint and using as reference material the very same figures I had used to make the complaint of being under supplied! The congressman evidently didn't read the letter and just passed it through. Really made a fellow feel good. It felt even better when some people jumped on our case and abused us after we came home.

Kontum, 5 August 1962

We have been living on a thread again with the very few planes coming through. The food situation looked pretty bad for a while, but an army Caribou finally came in, the hell with bad weather. There has been little or no mail.

Here is a story about the heroism of a young Vietnamese lieutenant. Two companies of Viet Cong ambushed a platoon of ARVN just outside Kontum not long ago. The ARVN troops were so heavily outnumbered it looked like a massacre was about to take place. It so happened that a young Vietnamese lieutenant in the column was riding in the turret of an armored car. This lieutenant sized up the situation and, barking commands to the driver and gunner, moved up and down the road firing viciously and skillfully at the Cong. The lieutenant died, mortally wounded. But the damage he did to the two enemy companies was fantastic. Single-handedly, so to speak, he shattered the ambush with his cool-headed command of the situation. He also saw to it that the two relieving companies of Civil Guard troops *(my later assignment)* were rapidly directed to the pursuit of the VC before he died. For two days the two Civil Guard companies chased the remnants of the two VC companies through the country around here in a series of running skirmishes. When it ended, barely a handful of the Reds escaped. Perhaps this explains why enemy road ambushes are not too effective around here. *(This incident is one of hundreds of similar incidents overlooked by the*

American elites [in the Kennedy administration] who both ran and evaluated the war. Think what could have been done by those of us in place at the time, had we been given the stuff and the leadership.)

The other day we were going up the road to our outpost at Tan Canh [Tân Cảnh]. Ed Mixon, Ed Opstad, and I were in the mess hall eating early lunch a few minutes before departure. In came Lt. Col. Sweet, the new boss. "They just ambushed the road," says he. "You better reconsider about going." Half of our party backed out immediately. But I thought a second. When the Cong springs an ambush, they run as fast as they can go, because it doesn't take long before air strikes and ARVN are all over the place. Therefore, the road is safer than ever immediately following an ambush [and definitely more safe] than it would be if it were several days since the last one, I reasoned. Not too many people were convinced with my thinking, which bothered me. The others should have told *me* this, not me them. *They* are the combat advisors.

Anyhow, I convinced enough people, and Mix and Ed and I roared up the road to Tan Canh and then returned safely. Each time I go up there I count the ambush sites, and each time there are more places to count. When we first started driving up there we kept weapons on SAFE. Now we leave safeties off and spray a few rounds at suspicious-looking bushes. Fresh mine holes pock the road, and men who used to ride the same route casually are now leaning forward, straining, with weapons at the ready. But as I told you before I left, I keep my eyes peeled on those sorts of things, and I know what to do and when to do it. *(We were professional soldiers and we knew how to exercise fire discipline so that we did not have any incidents of shooting up civilians by accident. But we had some close scrapes, such as the time I was surprised by a young Montagnard woman with a papoose on her back. To this day I still sweat about that one!)*

The VC was all around the road, yes. The fact that they had attacked had everyone scared to go near the road. But everyone overlooked the fact that though the Cong was about, they were making for the hills in full retreat. It seemed to me the safest time to slip through, and it was. I knew that if the VC dared to spring a second ambush on the same road a couple of hours later and catch us, they would be so strong that they could have nailed us in Kontum anyway. We spent some anxious but safe hours in our vehicles. I shall return to Tan Canh after the next ambush, unless I know the situation to be different.

The war is with us, but I am safe and there is not anything to worry about with me. I can pick my times to go out into it, and I pick the good times so far. Those who do get caught and get hurt are caught flat-footed; reaction is slow and confusion reigns. All of our guys who reacted violently to the Cong ambushes up here, charging right back at them with a storm of fire and grenades, made the enemy flee from the scene. This leads me to believe that they have let many of us pass by unmolested merely because we looked good and ready for a fight, with weapons and grenades at the ready and eyes searching everywhere. I've felt a couple of times the eerie feeling that I was rolling through a perilous VC ambush site with hostile eyes and weapons aimed at me. But my men and I looked so damned ready to strike back violently that they let us alone and waited for easier prey. I shall never know this for sure. *(I sounded confident here, but that didn't mean I was relaxed. At one point in time I went about 120 hours without sleep and had to have help from the doc to finally get knocked out and get some rest.)*

This is a very strange war. Many times you are almost sure you are heading into a fight. Others get into them, others get hurt. But I have gone through the emotion of a score of firefights and skirmishes, and either they chose not to tangle with my bunch or they were not there. It is hard for me to say that I am becoming a combat veteran, but then I can't say that I am not either. *(It was early; stuff happened later that made it an easier call.)* My mental attitude towards a fight and the attitude I demand from those who follow me into a potential fight is one of absolute viciousness. This is the only key to survival in this kind of war, because the Cong will never start a fight unless they feel sure they have you. And the only way out is the fiercest, most vicious counterattack. The Cong don't seem disposed to tangle with groups that come through them around here ready for a fight. I have seen Vietnamese and Americans even who go out on the road just as if they were driving from St. Charles to St. Louis. These are the ones I've also seen carried back in stretchers. The Old Man [Tencza] went down fighting, though. He was a victim of bad luck. They shot down his chopper and shot him as he got out. The major managed to get away. That was a rough and sad day. "Terrible Tony" made life tough, and he was a driver, but he was also affectionate like a father and a man of great strength and integrity. There were many of us who shed a tear for "Terrible Tony." The 15 of July 1962 was a rough day in Kontum, Vietnam. We lost many good men on that and succeeding days.

At this stage of the war in the mountain highlands, the enemy forces were made up of numerous small groups of "bandidos" and a few larger company- and battalion-sized units wandering about. The smaller units were not very good, but the larger units sometimes were. They were bolstered by supplies and personnel from the north. The larger units attempted to take on the ARVN from time to time, so they considered themselves decent troops, and they had the experience from the French war days to make a sound judgement. For us, this was a winnable situation at that moment. Too bad the war was being run by a small group of advisors to President Kennedy, none of whom had a handle on the situation, and all of whom apparently wouldn't listen to those of us in the field.)

Pleiku, 8 August 1962

Yesterday morning early, in the gloomy drizzling rain, we boarded an Otter (a small fixed-wing army aircraft [a U-1A]) at the Pleiku airstrip. After some delay we finally took off and circled up slowly through the clouds until at last we were above a sea of white foam heading east. For nearly an hour there was little to see except the clouds until we neared Ankhe [An Khê], along the Pleiku-Ankhe road. This was most interesting, since it was the site of the battle between the French and the Viet Minh described in Bernard Fall's *Street Without Joy*. Shortly thereafter we left the clouds and broke into a sunshiney scene. Dead ahead I could see the sparkling blue South China Sea, and all around us were the green-covered mountains and the prosperous-looking valleys. Our first stop was Quin Nhon [Quy Nhơn], which was truly a beautiful little seaside town. From there we flew on down the coast to Tuy Hoa [Tuy Hòa]. The landward side and the seaside reminded me so much of the California coast. The mountains had the same green look, and the sea that same delightful sparkling-blue look. We landed at Nha Trang late in the morning, and it just happened that my business there was finished very quickly. I got myself on a flight manifest on a Caribou (a two-engine army transport) to Pleiku in the morning and ran into a couple of chaplains. One, Father Grotevant, I had met in Kontum a few weeks ago when he was visiting. The two good fathers were also awaiting transportation. Since Nha Trang is Father Grotevant's home base, the three of us had a good time that evening. He drove us down the beach along a beautiful boulevard lined with palms and gracious French buildings on the land side and a beautiful sandy beach on the sea side. The American quarters where I stayed was a very fine former French hotel with a nice yard around it. It is the only building I've been in in Vietnam where you could actually use tap water safely to drink.

The two good priests and I jumped into bathing suits and whisked down to the beach for a before-supper dip in the ocean. This was really living for a change! The water was just the right temperature, and the force of the waves was so powerful that we had great fun trying to stay on our feet and riding on the crest of the tide. After an hour of this fun we went back to Pacific Hotel, showered, changed for dinner, and had a before-dinner drink in Father Grotevant's room. Then we ate a nice turkey dinner. After dinner I took a nice stroll through the hotel grounds and enjoyed the sunset. Strains of beautiful music filled the air, then later, a movie. *(This was one of the most beautiful places I have seen in my life. I wonder how it fared as the war progressed.)* Next morning, I took the Caribou to Pleiku and a HU-1 (Huey) to Kontum.

Kontum, 16 August 1962

The war out here continues on. I now know the feeling the ancient Romans must have had as the barbarian hordes swarmed over the frontier into the provinces and finally to the gates of Rome herself. Here the Communists pour over the border *(from Laos)*, fight flirtation skirmishes with the few Viet troops on the border, and then disappear into the endless mountain jungle. A couple of days later they fall upon a frontier outpost a little further inland, or perhaps clash with a roaming Ranger unit. Then we hear next that they have pushed on past us on the Ho Chi Minh Trail, or that they are marching perhaps on the provincial capital of Kontum itself. Then somehow we lose track of that group, and others moving in occupy our attention. They have threatened to attack and destroy my fort, and last week they even staged the forces to have a try at it. Captured documents revealed their intention to attack three objectives in the Kontum area. I am highly complimented that my fort was one of the three.

For some reason, perhaps the distance they would have to travel after the attack to escape retaliation, the fights did not develop, except in our minds and preparation. We were given ample warning, and we were waiting; we would be a tough nut to crack. It would take a large battalion to sweep into here, a major fighting force of the VC. And even if they got in, the fight would have just begun. They would have to pull out of the fight and get away well before daylight to escape undetected.

Once engaged in a fight, it is very hard and costly to break out of it unless both sides are equally willing to break it off. The enemy's risk in hitting us is great, and his knowledge of this fact must be preventing him from coming in.

Given the right conditions, a mere platoon (50 men) could wreak havoc amongst us. But if we get enough warning to prepare and to get to our positions, it would be a lulu of a fight. The VC can't afford to be beaten off, and they can't afford to overrun us and then get caught while escaping. These fears of the enemy keep us safe for the moment.

A number of enemy prisoners have been brought in here on hard-labor details. They are mostly Montagnards, whose allegiance to Communism is tied to their families being held as hostages by the Cong to assure their loyalty. *(We had reports from some of our advisors who pursued retiring enemy units and came upon a dead enemy machine gun crew shackled to their weapons and a tree.)* For this group, their forced loyalty was enough to make them fight government troops and get captured in the process. I need laborers around here, so ARVN sent me these prisoners for a labor force, and they are good and willing workers. They are a filthy bunch dressed in rags and suffering from malnutrition. All the Viets give them to eat is a couple of handfuls of rice a day served out on a dirty sheet of paper. I became aware of their plight when I saw that even three of them could not lift a load that was a toy for me. They didn't look undernourished, but they are. When my sergeants and I saw this we started looking for ways and means to feed these guys, but it wasn't much.

This afternoon I dug out some soap and made them all stand under a hose and wash up. The Viet guards do not seem to mind our doing this, but they don't seem to care what happens to these guys either. Such a case of utter misfortune I have seldom seen. The VC seize their families as hostages and force them to join up. Then they must go through the terrible rigors of a guerilla barely surviving, and bang, the government captures them. Since they are a minority group and considered VC, they get thrown into prison for hard labor and are all but starved to death. They have no place to turn and little hope. But damned if they still can't smile, and they don't seem to bear a grudge against anybody.

We Americans have developed a soft spot for our Montagnard "VC" laborers, and I'm sure our Viet compatriots, or some of them at least, think we are nuts. Some seem to sympathize and cooperate when we try to help these guys but would never think of originating anything decent themselves, or get the hint. I have read many noble-sounding words about how we must reach our hearts out to the people to win this sort of war. It is very difficult to find many cases where we have actually attempted to do so. One certainly feels swept along by the trend of events in a most abashed way at times. It is strange to see how I who am an American can have so much, but he who is a Montagnard has nothing and suffers everything in so far as the things of the world are concerned. But the spirit of the Montagnard is as noble as the finest man who ever lived. He is willing and loyal even to his oppressors; he seems to bear no grudge. He remains cheerful, accepts his lot as though he chose it himself, and asks nothing. What is more, he has a personality that makes him immediately liked by Americans. He is curious, friendly, and acts like "one of the boys." He is in the subtlest, simplest way a saint. He must undoubtedly be, deep down inside, the nearest thing to what God desires man to be. The plight of the Montagnard is one of the tragic but beautiful stories of this war.

This is a very strange war, and even the official reports amongst us are conflicting as to the progress of the thing. The friendly governments in their press releases are striving to show success in the number of swift-striking actions they launch against the VC, and also the increased effectiveness of our forces, the reduction in reaction time to strike back, and in the number of the enemy we claim we kill. The enemy feeds more and more troops into the south, far more than either side will publicize.[1] A vast number of these troops are passing right under our noses here in Kontum, streaming out of Laos. But it is bad publicity for our side to publish this, since it would hint strongly at the gross error we made in the Laotian settlement. Because of generally poor communications, I doubt if the Red brass in Hanoi really knows how things are going on their side. I doubt if anyone knows. They just keep pouring in VC units, control and supply them as well as they can, and hope for the best.

If anyone really knows what is going on, I doubt it for several reasons. Mainly because the top dogs in this fight seem to believe what they want to believe, and they see things in that light only. The Reds display the better top-level leadership, but they work under the tremendous handicap of the type of war they have to fight and the probable lack of

---

1. See chapter 10, "This Intelligence Problem Must Be Solved," and chapter 13, "Secrets in Saigon," in John M. Newman, *JFK and Vietnam: Deception, Intrigue, and the Struggle for Power (1992).*

information they should have. *(Remember, reader, this letter was written by a young paratrooper captain in his middle twenties actually on the battlefield in mid-1962 and who, a year and a half later, lost his career and veterans' benefits trying to set things right by saying, "Hey! Will somebody please listen? We have got to stop the way this thing is going before we are here ten years and lose fifty thousand men! Doesn't anybody care?" Most didn't care until they took their tourist trip to The Wall decades later.)*

The Viets definitely are the sort who believe only what *they* want to believe, and when things continue to go awry, they accept it with their typical stoicism and repeat the same mistake the next time. They do not wish to change attitudes and methods merely for the sake of winning the war, and they are coming around much too slowly. The Americans approached the problem by throwing money, materials, and men at it. While the people back home who are tied up in this thing hoped that would make the VC bogeyman go away, those of us over here find ourselves so entangled in our own resources that most of us are spending our time just trying to perpetrate and maintain our own monster. So we are in this thing mainly concerned with ourselves, and very few of us are actually helping the Viets in the war, although we are all stuck in the middle of it.

Since communications are so poor in Vietnam, I doubt if anyone on either side really knows what's going on. *(What would we have thought if we knew then the information that recent books have revealed? If you read Newman, you would see that General Harkins and Colonel Winterbottom, the proteges of General Taylor, were apparently deliberately reporting downsized enemy order of battle intelligence for their agenda. If you read Newman, Prouty, McMaster, and Winters, you would see that the primary focus of the top administration people was the 1964 election, and the war was going to be run with that in view as the top priority, no matter who was hurt.[2] If you read Hersh,[3] you could catch a line of thought that the*

---

2. In September 1963, Francis X. Winters reported that Kennedy told journalist Charles Bartlett he was planning on quitting Vietnam after his reelection. "We have no future in Vietnam," Kennedy said. "They're going to kick our asses out of there. I can't give up before 1964. I couldn't go out there and ask for reelection after giving up two pieces of territory (Laos and Vietnam)." (Winters, *The Year of the Hare: America in Vietnam, January 25, 1963–February 15, 1964* (1997), 192.) Note, dear reader, that they were not willing to give up the 1964 election, but they certainly were willing to give up the lives of my comrades in arms who fell during that time to cover their politics.

3. Editor's note: Probably investigative journalist Seymour Hersh.

day-to-day private immorality of the top people in the administration was fouling up their ability to run the country and its war. If you just look at the results, it doesn't add up that such an arrogant bunch of elitists could have engineered such a stupid result, so there must be something to what these authors are saying.)

I think, as the guesstimate of one intimately involved, that this war is far bigger than people think. I think the enemy is much stronger than commonly thought. The reason we are killing more and more of them is not necessarily an indication of greater efficiency on our part but rather that there are more to kill. The enemy's efforts do not appear to be seriously hampered in a region where the government makes a big kill. The hugeness of the war can be illustrated in the fact that for the many reports of successful operations that are published, there must be 10 more reports of unsuccessful operations. And each of these operations, which are going on constantly, involve many, many troops on both sides.

This is a strange, obscure situation which has little reliable information to offer, both by the character and type of the war and country, and also by design of both sides involved. *(Boy! I didn't know how right I was on that one, and how criminal the problem.)* The winner, to me, appears to be the side that will stick it out the longest, rather than any brilliant generalship. *(Target!)* Our side, if lacking anything, is certainly lacking leadership at the top. *(Direct hit!)* Red leadership is better, but they cannot pour in the resources *(at least in mid-1962)*. I expect to be long gone from here before either side breaks. We won't pull out, because, as I can see from the nature of resources we are dumping in here for our own use, we hardly can. As long as we stay, some sort of Viet government will stay propped up under our wing. On the other hand, Red tenacity is going to have to be tempered, and that is not likely before they quit. Of course, a break could happen on either side to change the entire picture. *(President Diem's assassination changed things fifteen months later.)*

We are beginning to win some telling victories around here. Two weeks ago a large VC force stormed a Montagnard village in our area, with the idea of sweeping through onto Kontum. The Montagnards beat the hell out of them, and the Cong had to leave behind 67 bodies. This was an indication that they lost many more because they *never* leave dead or wounded behind. When they do, it means they took a real beating. A couple of other places have held out successfully since then, inflicting heavy casualties on the Cong. This week the 3rd Battalion of the ARVN

40th Infantry caught a VC unit, killed 1, wounded 4, and captured 36 with many weapons. These things mean that we are beating the VC in Kontum Province at last. Let us pray that we can keep it up and it spreads to the rest of the country.

Kontum, 20 August 1962

This year so far it appears we have the VC in Kontum Province in trouble. Past years this time they were over-running towns and forts like the gates were left open. A few of the usual "pushover" places beat the hell out of them in a surprise turnabout, and our guys here have been chasing them around enough that most of their plans never get a chance to get underway. The roads are even safer now, as they seem to be so off balance that they have not been able to set ambushes. The last ambush I can recall just ahead of us was the day I drove up to Tan Canh. We raised so much hell then that they never did come back yet. We've been getting some air fighter cover lately too, and the VC are scared of that. *(These were a couple of A-1E Skyraiders. For some reason we called them AD6s, prop jobs.)* I wouldn't like it either if a pair of those fighters screamed down on me raining all sorts of destruction. (The napalm firebombs were real peace makers.) There seems to be much more of a spirit of offensive action about.

Almost nightly, artillery near here is firing and troops are prowling in the nearby jungle. Maybe our side is on the move. We will see in the next few months whether we can finish it now or not. I am getting anxious to get out of this compound and get with the pursuing troops. I would like to help mastermind some of these traps they keep trying to set for the Cong. They rarely catch them, and I've got a hat full of ideas how. Maybe one might work.

We have seen many clippings and heard much from the States about Col. Tencza's death. Much of it is blown-up newspaper stuff. But he did have a nice military funeral at Arlington, I see. That was a rough time, but thank God no more of that has been our lot yet. We had a humorous scare the other night. We heard crashes all around us and we could have sworn the VC had set up a mortar and were popping at us. I tumbled into a foxhole at 2:30 a.m. and had much company. After we learned it was friendly stuff being fired at the Cong, we "veterans" sheepishly returned to sleep. Older Korean War vets were fooled too. Anyway, a good laugh at ourselves.

Kontum, 10 September 1962

We had a minor accident today. I was teaching one of my Vietnamese sergeants how to drive a jeep, and we rammed head on into the front gate of
the fort. As Thach roared toward the gate on a definite collision course,
I hollered "NÍN LAI!! ARRETEZ!! HALT!! STOP!!" in four different
languages. Panic-stricken, Thach slammed the accelerator to the floor
instead of the brake. It all happened too fast for me to dive for the brake,
but the guards fled for their lives, and all I could do was brace myself and
keep roaring "NÍN LAI!! ARRETEZ!! HALT!! STOP!!" I was still
hollering after the crash, and then as I recovered from the impact—me
still hollering, "STOP IT!! CUT IT OUT!!," Thach threw it in first in
a complete panic, his eyes wide with terror, and rammed the gate once
again. It made a big noise and showered debris all over the place. It was
also quite a spot for a public spectacle. Poor Sgt. Thach looked like he
wished he could evaporate, and everyone around descended on the scene
with howls of glee. I must admit I was laughing myself when I saw the
damage was slight. A nearby tree lost a limb and many leaves, but for
the life of me I don't know how the tree got into the act. It was immediately obvious, once I saw that we were all in one piece, that I get Thach
and the jeep and me out of the vicinity as soon as possible. So I backed
off the jeep, found it would run, and scooted off behind the motor pool
shed to hide the evidence, with Thach running along behind me. The
front bumper of the jeep was banged in, but for all the crash and commo-
tion, that was the only damage to jeep or gate. What a moment! What
a noise! I thought we had destroyed everything! I was in great fear the
colonel would be around, and that didn't slow me down any, I assure you.
I moved as fast as I could and still maintain dignity. My friend Lt. Ming
over at the ARVN Ordnance Company said he'll fix it up for me and it
won't cost anything. That may end the driving lessons for a while. I don't
think Sgt. Thach will want to look at a jeep if I want to teach him. Loss
of face for both of us today! But one good laugh.

Tan Canh, 13 September 1962

Fighting has stepped up around here, and it's hard to tell who is winning.
The VC has stopped interdicting the local roads lately, and I hope they
don't plan anything as long as I am moving around on them. I think they
may be losing some of the popular support they need and at the same
time are calling their small bands together to form larger units.

At any rate, the scope of engagements has grown larger. Battalions and regiments are now squaring off in the jungle, and nobody really can assess the outcome. This bunch I am with now, the 40th Regiment, launched an operation against the Laotian border last week. On the way there, one of the battalions met head on with an equally surprised but larger Viet Cong unit. A fierce fight developed. All indications are that the VC thought they merely hit a platoon-sized patrol. So instead of running, the VC attacked. Opposing lines closed to 30 meters at several points, and the battalion from the 40th was completely surrounded by the enemy. In two companies all the Vietnamese officers were killed, and an American captain, Roy Muth, and a Lt. Joe Hicks were the only officers around. The company commander of one company was rushing forward to emplace a machine gun with Roy Muth when he was killed.

Bullets were tearing up the jungle and raining debris everywhere. Joe Hicks said he couldn't get close enough to the ground because his buttons were in the way. The enemy made fanatical attempts to retrieve their own casualties. When a Cong was hit and fell, a recovery party would rush out and recover his body. This must be because they don't want us to be able to tell the world, "We are tearing up the VC." And they don't want North Vietnamese regulars identified as such. About the time the situation looked critical, the other battalions of the 40th moved up and the VC broke off the action and flat disappeared. The 40th proceeded westward and captured vast amounts of VC food supplies before returning. *(The scope of this fight was probably a little larger than the Battle of King's Mountain in the American Revolution, and it took place in even wilder and more remote country. The fight had no known significance other than to attrite (or bleed) both sides. To my knowledge, these kinds of fights were not reported by anybody outside of the after-action reports written in the letters of the participants. Great opportunities for learning were lost here and at many times during this period, but our leadership had other interests, as you can read in the books by Prouty, Newman, and Hersh.)*

Vietnamese artillery fire is improving greatly as well. Villagers from a remote village tell of an enemy unit that was *wearing South Vietnamese Army uniforms* with red scarves. Suddenly, artillery fire began falling in their midst from somewhere. So great was their panic that they started firing amongst themselves. Officers shot many who tried to run away.

In another engagement in which ARVN artillery fired, 20 VC bodies were counted on the spot, and Montagnards reported that there wasn't an enemy soldier who walked out who was not carrying somebody else.

The other night they (the VC) attacked an engineer platoon of our division engineer battalion. The enemy raced headlong into a minefield, and their losses were heavy; they panicked and couldn't get away without a severe beating.

Last night they hit the outpost of Dak Mot [Đăk Mot], which I visited today. The Dak Mot outpost is on high ground overlooking a river bridge. On each side of the bridge are two smaller fortifications which are supported by the larger fort of Dak Mot. Dak Mot is commanded by a green second lieutenant who, until last night, had not yet been in combat. A large enemy unit stalked the main fort on the hill and completely surrounded the place. The enemy threw in grappling hooks and pulled down sections of the outer bamboo barricade. The fort's lone 60mm mortar fired up a storm, and sporadic small arms fire developed. The lieutenant had telephone contact with regiment here at Tan Canh, and he was in a very shook-up state. The enemy was about to pull down his outer barricade and rush his inner defenses. Cecil Henry, who was over there at regiment, hollered at the exec *(the Regimental Executive Officer, the second-in-command of the 40th)*, "Why don't you fire the artillery?" The CO *(commanding officer)* was out meanwhile dispatching orders to the 155 howitzers at Ben Het to fire on Dak Mot. No one knew how long this would take. In the meantime, the 105 howitzers at Tan Canh sat silent and the enemy was pulling down more and more of the outer barricade of Dak Mot. In desperation, Hank turned to the Regimental S-4 *(supply staff officer)*, a man with a reputation for action, and pleaded with him. The 4 picked up the phone to the guns, and 2 1/2 minutes later the landscape lit up and shook as the guns roared out. The enemy apparently did not know that we had the guns within range. At the same time, the Ben Het guns commenced firing. The ground around Dak Mot was severely pounded. The enemy withdrew and fled.

I have news from further south in our division sector that the 42nd *(ARVN Infantry Regiment)* has fought 5 engagements in the past 36 hours with a large enemy force. Our casualties are reported not too heavy.

Things are picking up, as you can see. We are getting out and catching them and making them fight. When they attack us, we are blasting them and holding. We are giving them hell. Their desertions are increasing. We *may* be winning, but in the jungle and uncharted terrain it is rough to realistically assess anything.

I now have 3 American detachments under my control. Kontum, the one at Tan Canh, and a new one at Cheo Reo [now Ayun Pa]. The high-command types in Saigon tell me I'm one of the five largest operations in Vietnam. I am the youngest captain in the theater with such responsibility *(for whatever that's worth).*

Kontum, 18 September 1962

We had another helicopter shot down by the enemy today. But we were lucky. They were all hurt seriously and we got them out before the enemy got to them. They will all heal in great shape; they took a real beating but beyond doubt they will all survive. Enemy fire apparently knocked them down, and the enemy continued firing after they were down. As they were picked up, the enemy could be heard moving around in the underbrush trying to get to them first. It just luckily happened that in Kontum we had a C-123 and a C-47 sitting on the ground at the time. We held them up and shipped the boys straight to the hospital on them. [It was smooth] like clockwork, but we unfortunately are getting to be old hands at this sort of thing now. Captains Wilson and Sanders, Sgt. Morton, and Mr. Dennis, a civilian radio expert (CIA?), were the Americans wounded. Captain Taum, our Viet vice governor, whom I've mentioned before, suffered a broken leg and arm.

That runs up our toll somewhat. It is difficult to believe, but we've probably got the highest casualty list in Vietnam.

Today I was presented with a gift which I didn't especially want. A stuffed wildcat *a la* Vietnam. I saw one someplace this morning and asked this Vietnamese officer, "What is that?" "Oh ho ho," he hollers, "I get one for you!" I hoped he would forget about it, but not thirty minutes later he returned in his jeep and presented me with this stuffed wildcat. I pretended as though it would be my most prized possession, and now it sits in my command post glaring at whoever comes in. Our mascot, a cute young pup named Fluffy, always trots in and sleeps under my desk between escapades. Fluff trotted in, took one look at the stuffed cat glaring down at her, and found a new place to rest.

Here are some stories about my encounters with wildlife during this period: One afternoon, Sgt. Thach found me and asked me to come to the gate area;

we had visitors. We walked over there and found a scene of much babble going on in at least four languages. Sister Marie Louise and some of her sisters were there with their van and a small group of Montagnards. On the ground beside the van was a long bamboo cage containing one very angry thirty-foot python. The huge snake was doubled up in the makeshift travel cage with his mouth tied shut. One look in the snake's eyes and you could see he was plenty mad. In the meantime, an American sergeant, some Vietnamese and Montagnard soldiers, and Sister Marie Louise and her party were all jabbering away at the same time. As soon as I arrived on the scene they all descended on me.

"You buy snake $20 American, now!" The enterprising Montagnards were looking to expand their profitable crossbows-and-artifacts business by expanding their line.

They added a benefit to the sales pitch: "Snake eat rats!"

I countered, "My Montagnard guard platoon eats rats!"

They countered with something like they'd brought this new line of merchandise all the way down here to Kontum, and Sister was kind enough to pick them up and give them a ride in. At that point Sister Marie Louise went into a sixty-mile-an-hour French pitch with much gesticulation. I let her run down and then used my best one year of UW French to tell her, "No deal." Many disappointed looks were shared, and they hoisted the angry snake up to the top of the van and headed for the Kontum marketplace.

I wasn't kidding about the guard platoon eating rats, though. One day early on in the command post, I reached down to open a desk drawer and right next to my hand was a huge rat the size of a small cat. It snarled and leapt about a foot in the air. I jumped about two feet and somehow slammed a wastebasket down over the rat. Then I called for the guard to dispose of it, and two of them came in, followed by the ever-present Sgt. Thach. They peeked under the wastebasket and their eyes lit up. They flipped the wastebasket over, covered it up, and headed for their mess hall chattering, "Ngon quá!" (Very delicious.)

Another evening in the officers' lounge, one of my compadres spied a rat running down the wall, drew his .45 pistol, and BLAM! It almost caused an alert, as all those not present started to grab weapons and head for the firing positions. After that, I decided it would be undignified to shoot something unless it was really big.

Vietnamese kept pythons as pets. Some of my advisor buddies had a Viet counterpart whose home they visited in Kontum City. They had a big python that would drape itself over the fireplace mantel and take a nap. One blistering hot day our two guys drove up in their jeep to visit. The big python was in a tree outside the house and apparently thought it would be nice to greet the guys. The thing flopped down out of the tree onto the jeep hood, which was as hot as a frying pan. Writhing snake and both Yanks flew off in three different directions in world-record time.

I woke up one night to the sound of an aircraft flying around overhead, apparently lost, and I also heard an unfamiliar motor sound, definitely not one of ours. Soldiers in combat get to know all the sounds; they could be important to continued good health, as they can tell you a lot about what is going on around you. This stranger was flying around in the night, obviously lost, I thought, and we were told in the morning it was an enemy aircraft and was shot down by VNAF (the South Vietnam Air Force).[4]

Kontum, 24 October 1962

Hou Ban [Hòa Bình, part of the town of Ayun Pa], more commonly known as Cheo Reo, is located south-southeast of Pleiku. Pleiku is directly south of Kontum by about 40 kilometers. Many, many VC down there. If you could see this place now compared to when I arrived, you would see that we have made it into a real fortress. There is much to be done, but under normal conditions it is doubtful that even a VC regiment could successfully attack here now. But I do not forget the many incidents of history where a handful of men surprised and captured a place this strong. The essence of our strength lies in our alertness. Many days and nights we have scrambled into our defense positions on practice alerts, and we keep some men on guard at all times.

I had several months of worry because the VC were strong enough to tackle us there for a while, and our defenses were not too adequate. A mere enemy company could have clobbered us, and our warning and protection systems were hardly good enough to get us out of bed before the enemy was streaming over the walls. Now we have two different types of barbed wire entanglements several meters thick about 100 yards out in the most likely places. Also a triple concertina fence and a box fence

---

4. William J. Rust, *Kennedy in Vietnam: American Vietnam Policy, 1960–1963* (1985), 208. In the second week of March 1962, President Diem charged that Soviet planes—which were being flown by Chinese and Vietnamese pilots—were supplying the Viet Cong.

with spike mines on the outside. The grass is kept short and all under-brush is cleared away so we can have good fields of fire. We have artillery concentrations and barrages plotted everywhere. I tore down old forti-fications and put in new, larger, stronger ones where there were not any. We improved our communications and command control around the perimeter. We have made our soldiers much more alert than they were. We have increased our firepower tremendously. We have 6 machine guns, 4 BARs *(Browning Automatic Rifles)*, 30 submachine guns, 60 carbines, and 30 M1 rifles with thousands of rounds of ammo and hundreds of grenades. I estimate we should have about 6 minutes to react if we get hit. The barbed wire and the guard pickets will give us that time. That's one minute more than we need to start bringing effective fire on an enemy trying to sneak or storm in here. What we have now appears too strong for what the enemy is willing or capable of doing now. BUT the best pol-icy is to go on living thinking that they just might try it. That's our best insurance toward going on living, period. I have become a pretty fair fort builder for a soldier who is offensively minded. *(A couple of times, enemy leaders sent in notes telling us that in honor of some Communist holiday or other they were going to overrun our place. I sent back word, "Try us out!" They never did while I was there, but I believe in February '65 they did attack and got pretty far in before they were thrown back out. I suspect our successors were not as careful as we were.)*

When my people and I get into a scrap, and there have been some minor ones, the enemy is seeking a fight because he is stronger than us. When I run our supply convoy up the road to Tan Canh, I go far more apprehensively than I would with a battalion of first-rate troops. This is because I am taking an ambush target up the road: Slow, unwieldy trucks loaded with valuable supplies, often ammunition and gasoline. Only a few guards made up of troops not good enough to be with line battal-ions. Up the road we go at 20 miles per hour through the heart of enemy territory. In this situation you are sitting ducks. The enemy can get in the first and biggest punch, and you are confronted with a nightmare (and possibly only a few seconds of remaining life). You know in the first place they would never tangle with you unless they knew they had you. You get into a fight while with a battalion and you know the VC only want to get away. The fact that you caught them in that situation means that you have the advantage. The fight takes place in a location where you have plenty of cover and concealment and many staunch comrades on all sides.

You are with a unit that has the punch to do the job. You are not stuck in the road a sitting duck. So don't feel bad about my wanting a fighting battalion. Very few of those guys *(American advisors)* get hurt in comparison. Besides that, you are much more mentally able to cope with various situations. Since August (1962) there has been little trouble on the roads around here. This is because the battalions are chasing better.

Kontum, 10 October 1962

Major Riley, Regimental Advisor of the ARVN 41st Infantry, went down to Hou Ban yesterday to try to talk Elmer Biggs into coming back and taking over here so I can take the 1st Battalion. If he brings Elmer back, I will join the battalion as soon as I break in Elmer. The 41st people are anxious to get me into the field as soon as possible so that I can take the 1st Battalion through its next big operation, which may last through Christmas. Phu Ban Province [Phú-bổn] is more the enemy's than ours, and the battalion has to go out and destroy the enemy and secure it. The big hitch will be catching them and getting them to stand and fight. Actually, the trick is to catch them in a trap so they have to fight. This really has me interested, and I sure do hope Elmer comes back with Major Riley. Riley seems very interested in making this change at this time. I suspect it is because he feels I will be able to take it better than Elmer, who is much older. *(Events didn't work out on that one.)*

13 October 1962

The boys from the 20th Ranger Battalion moved into a Viet Cong "safe area" and came upon some rice fields, truck farms, and livestock farms that the enemy was using to feed their troops with. A Viet Cong unit attempted to put up a stiff resistance. One company of the 20th kept maneuvering around to the rear of the enemy in an attempt to envelope them. Each time, the VC fell back in an orderly fashion and continued resistance. Finally, the fight obviously lost, the enemy broke contact and disappeared. The farm was secured by the 20th, and about 30 enemy dead were counted. Much livestock, several villages, and a great deal of enemy crops were captured. The Rangers destroyed what they couldn't use or carry away. This is an example of the success we are beginning to have. The enemy, countrywide, has been losing about 3 casualties for each 1 of ours. The tide is apparently turning. *(As I transcribe this in 1997 and consider the now-declassified historical documents I've been able to read in the last few years, I wonder where was the Kennedy administration and the Saigon MACV bureaucracy? We had a chance! We had a chance! They blew it, and they covered their tracks by classifying the records for years. The uninformed public still thinks they are wonderful.)*

Nha Trang, 16 October 1962

Well, here I am in this beautiful resort area living it up for a change. I came down here yesterday to this picturesque French resort town to go to our hospital down here and have a little minor surgery done for piles. These are easy to pick up over here, and many guys have had the same problems.

Anyhow, I went over to the hospital this morning and was cut and burned on some and then released and told to return in a month for another go at it. I then discovered I could not fly back to Kontum until Thursday a.m., so I am in a position to "convalesce" most graciously.

Last night when I breezed in here I reported to the MAAG detachment HQ at the plush Pacific Hotel and found an old acquaintance, Major Paris, sitting in the same seat I sit in at Kontum. The major really saw to it that I had a place to stay. He moved me into Villa Saint Paul [Saint Paul Hotel] on the seashore. Another guy from Kontum, Frank Ianni, and I are sharing this gracious place. Undoubtedly it was once the villa of a high-ranking French officer. Today it is mine. The place is surrounded by a big stone wall and has spacious grounds covered with walks, gardens, arches, alcoves, and trees and bushes. Beautiful palm trees everywhere. The villa itself has about 5 rooms with a little servants' cottage attached to the rear. Also within the grounds is a slightly larger villa, Villa Neptune, occupied by an American major and his family. Are they living it up compared to the majority of us over here! A small street runs from the two villas through the grounds to the front gate, across the picturesque boulevard outside, through the narrow little beach park, and to the beach itself. You could hop in a jeep and drive the 200–300 yards right down to the sea if you wanted. One can hear the waves of the sea from the villa, and from the upstairs balcony you can see out to the water.

The view down on the beach is magnificent. To the north and south several miles apart are two peninsulas that gracefully dip into the sea, with beautiful old French buildings on them. Beyond the northern peninsula is a still much larger peninsula of towering mountains which encircle the town all away around to the west and south. They remind me much of both California and Hawaii. Out in the bay itself are four more mountains rising strikingly out of the sea with great majesty and beauty. They resemble mountains much more than islands, and they seem to be uninhabited.

The seascape is beautiful. The boulevard is graceful, well paved, and thoroughly French. On the west side are the villas, schools, church buildings, and hotels, each enclosed in the quiet solitude of its own wall and grounds. Between the boulevard and the clean, sandy beach is about a 100-foot-wide long, narrow park lined with sheltering palms. Beyond, the sparkling sea, the mountain islands, the fishing boats, the horizon, and the sky. The scene is especially beautiful during sunrise [and] at night with the moon, stars, and clouds. It is very nice and restful to be here. I think I have had about 3 breaks from the job since I have been here. I was here once before overnight and had a chance for a before-supper dip, and I had a few days in Saigon once. That makes a total of 6 days off in 7 months. No weekends off, and all workdays, which virtually average about 14 hours a day, many more, a few less. A simple thing like a day or two off in a place like this is a big thing in our lives over here. Here you feel free—just plain free, and it's great. In Kontum I feel locked in a prison, surrounded by mountains, jungle, the enemy, circumstances, the barbed wire of my compound. Everywhere you go you take weapons and a security escort. I don't so much dislike the place as the job.

The weather in Kontum is changing. The monsoon, a very mild one this year, has ended and we have extreme heat from mid-morning to early evenings, sometimes till late evenings, and then around midnight it gets downright chilly enough to use a blanket or two. Nha Trang stays hot but has frequent refreshing rains and breezes off the sea.

My spoken Vietnamese doesn't improve or lose ground. I gain a few new words, forget a few old ones. My understanding of the local natives speaking to me is improving, but not good yet. I can, however, say most anything I want to them. They say my pronunciation is excellent. This is good. I oftentimes go long stretches of a day in Vietnamese and then find myself getting tongue-tied when I switch back into English. I will speak good Viet and then return to English with a pig dialect: "I want—you go—you know?" "Same-same" and all that stuff. Many comrades get into the same trouble whether or not they speak Viet *(because many Viets spoke English like that)*, and we all agree with a laugh that when we return to the States, at first you will all think we are nuts with that Tarzan talk.

7 November 1962

I came over here with the attitude of a Crusader, and not a childlike one either, and now I feel more like a dupe in farce. All in all I don't feel too satisfied about this thing.

Saigon, 9 November 1962

The Cuban [Missile] Crisis appears to have enhanced our prestige over here. The only results we have noticed thus far is that some of the supplies we were hoping for evidently got earmarked the other way, and we shall have to do without for a while.

I have been lobbying lately to get back out with the field troops again. The same outfit is going to bat for me once again, 1st Battalion, 40th ARVN Infantry. They are out crawling up and down jungle-covered mountainsides chasing the VC at the moment and for the next 4 months or so. They made a nice trap last week, probably hitting about a company['s worth] of the enemy. The Cong left behind many casualties and 5 prisoners, and my prospective battalion had one guy twist his ankle on a jungle vine while chasing the so-and-sos. The site of operations is to the south and east of Pleiku. If this deal works out for me, I shall probably be joining them by helicopter when you next hear from me. It shall be miserable, but it's what I want to do, so that's that.

I have really got the old fort strengthened by now. Some say there isn't a VC organization large enough to carry the place, under the right conditions. I regard such statements as foolish and keep throwing up barbed wire and barricades and having the gun positions and guards checked. I remember too many tales about so-called impregnable fortresses where a small band of determined types carried the place. We sound the alert siren for many a practice at odd hours to keep ourselves on our toes. I feel we are safe and secure, but I will never let my boys hear it from me. I have kept myself and my people alert, preparing, and ready, and this seems a good formula, because the only enemy soldiers I've seen lately have been prisoners. In my present job, that's delightful.

The flow of mail from all you home folks has been the greatest single good thing I have over here. We are all the same way. The most often heard question that plagues me is "Did any mail come in today?" My mail officer (our signal officer), Lt. Sniadajewski ("Ski" for short, also a south Milwaukee Polack), finally solved this problem by erecting a huge billboard-size sign where all can see it that is painted bright red with black letters MAIL on it. When that goes up it doesn't take long for the word to get around. I will bet that thing confounds enemy observers out there in the bush. They probably think it's some kind of alert signal we use, from all the activity they see whenever it goes up.

Speaking of Ski, one night I was prowling around while most of the guys were watching a movie in the EM [enlisted men's] mess hall. I came upon Ski's sleeping quarters and whoa! A huge tarantula the size of a small plate was sitting on his door. Those things can jump five feet, so I stood there wondering what to do. I could shoot it, but that would start an alert, not a very captain-like thing. So I went down to my place and came back with a stick the size of a ball bat and decide to whack the thing. I lined up like a batter and swung, WHAM! I must have missed; the thing disappeared, was nowhere in sight. So I went down to the EM mess hall and quietly approached Ski.

"Ski, I got some bad news."

"What, sir?" Ski says.

"You've got a visitor in your quarters," I said.

A group of us went up and searched the area, but the monster had disappeared. I had nightmares about another Polack falling victim in the jungle, this time to a Hollywood monster of a spider. Uggh.

A sad incident happened somewhere around this time. It was getting on into the night, and two of our NCOs were arguing in the EM mess at the bar, something about the relative merits of two types of dogs. One of them got up and left, reappearing shortly with an automatic weapon. Suddenly he started shooting up the place, miraculously missing the diving men. Then he went out to his hootch (sleeping place) and barricaded himself in there. We started an alert, thinking we were under attack. Major Murphy, who was near the scene, faced him down and got him to surrender by the time I got there. The man was sent out for medical care, and Major Murphy went up in all of our estimations as one heroic officer.

12 November 1962

I really got a laugh the other day. I got a cold, panicky sweat come over me as I thought of Peggy *(my wife)* driving Michele *(my new daughter, whom I had never seen)* on the Mark Twain Expressway *(I-170)*. If you could have seen me you would have laughed in comic relief too. There I was in steel helmet with submachine gun, hand grenades, and an alert bunch of Viet troops ready for a fight about to hazard our way down a road which the enemy has been ambushing lately. And I'm having apprehensions about my wife's safety. I get scared but really don't sweat these circumstances too much. We have a saying over here which doesn't seem to offend the natives, and in fact they embroider it for us and display it: "When I die, I'm going to heaven because I'm already serving my time in hell." I suffer on-the-spot scares but do not suffer any growing case of the shakes. God will take me down when He wants to, and there is no sense in my making a personal mental case over it. Moments come and go just like that, and we don't dwell on it. You cross the land which has been a battleground since the French were here, your weapons at the ready. You anticipate at any moment the withering blast of bullets that caught so many before at the same places. You experience what you think might be the closing moments of life many times over. But you never lose your sense of humor, you are always ready to go out, you are ready for whatever comes. I do not basically fear these risks. But when I leave here and come home I know I shall have great happiness, and if I don't I know I shall have an eternity of happiness with He Whom I love most. Besides, I am an expert infantryman with much doughboy sense, and I am always alert and instinctively keep myself prepared for any eventuality. As we move down the road I am always thinking a couple of situations ahead. I can find nothing for myself or those who love me to fear for me.

Perhaps one reason why I come out of these touchy moves better than my French counterpart of 10 years ago is this. My French counterpart would not pursue an enemy who attacked him. We will. I will leave the vehicles upon the road and chase the so-and-sos for miles through the brush as long as I have so much as a squad left to follow me, until the line troops can take over. They know this, I guess, and they don't bother us too much.

Twice the Cong has sent word in that they were going to knock over my fort on a couple of their Commie holidays. They even assembled and staged the troops for it. Both times the Viets scattered them before they could make the final march. Things like this have me believing we are winning this thing. These things are very significant in a soldier's mind. I often wonder how much more the enemy can take. They are a dogged bunch to keep coming back under the conditions they must operate under. But we have hardly begun to tap our store of vim and vigor. I think we are winning a rugged battle of sheer guts and endurance.

Maybe the enemy is beginning to crack. I am almost convinced in my mind that Hanoi has lost control of the war. The Reds send so many units in here they never hear from again. I don't think anyone really knows what is going on anymore, but I think we know a lot more than they do, and enough to eventually win. I have lost a lot of respect for the enemy's highly vaunted intelligence setup. A year ago they were knocking over the outposts and villages around here one after the other. Now they are being beaten back from one after the other; they have not attempted to knock over anything, tough, and they are now and then getting caught and beat up in the woods. This is a war that resembles our French and Indian War, and it has a story that you folks back home are not getting. We are reliving the era 1757 around here. *(Top government leadership never showed any appreciation of the situation; while concerned about other things, opportunities slipped away.)*

I have a goodly number of fellow Wisconsin men and even more Big Tenners. The football news *(1962 season)* is providing a good deal of enjoyment for at least a dozen guys. I got the Iowa game *(probably shortwave)*. We are all having our spirits uplifted by the Packers and the Badgers. The Packer machine is really impressive. I hope they can go unbeaten. *(They didn't, but they did build on the Lombardi legend.)*

Kontum, 18 November 1962

We are working up some big programs for the kids in the area here for Christmas, which we will all enjoy. The kids like us. They mob us everywhere we go. For "ugly Americans," we have enough popularity with a lot of these people.

Kontum, Thanksgiving 1962

We started off with an early morning Mass said by Father Lange, an English-speaking French priest who gave a wonderful sermon about our American Thanksgiving. He seemed to know more about that than we Americans. It was delightful. He also came to our Thanksgiving dinner, which was a nice affair. The men had decorations everywhere. We had recorded music, waiters dashing around the tables and all. We even found an American flag to put up behind the colonel's seat, which was really nice. Most of us have not seen Old Glory since we came to Vietnam. It is amazing how beautiful our flag looked to us. It is hard for anyone to understand who is not here. We are American soldiers over here. But we are in this war under the Vietnamese national flag. We have not seen our own flag for months. People back home just wouldn't understand. But our Thanksgiving prayer was a prayer of thanks for our flag. It represents everything we have and believe in. It is strong and beautiful. The French priest told us that the world depended upon American soldiers, that we had a special duty in service to God. It is good that we have a day to remind ourselves of this. *(Little did we know at the time the abuse that was waiting for some of us when we got home.)*

28 November 1962

I did a lot of reading and studying going back to the fifties to prepare myself for this, to make the best possible contribution. I practically wasted my time in a sense. At any rate I suffered enough of a jolt that I have gone from being a man of complete dedication to something else completely. I am not completely bitter, and although I have had many bad moments I have stuck with my faith. My main interest is no longer the army, because I feel betrayed by the army. It is now my family and myself that I am concerned with. There are a few good reasons why I plug away here. The best of these is I suppose that it's God's will that I endure this well, and therefore I must. Then, too, I have been a soldier too long not to. Also, I have my own feelings of personal satisfaction to work for. And it is obvious that I can and have learned a great deal. *(It was like being in a prison under sentence of possible death for crimes you didn't commit.)* I am a little bitter about the time I spent working and studying when I could have been paying more attention to my wife.

Kontum, 2 December 1962

We have had a tropical storm close to Vietnam this past week which has not affected us too much except for some high winds and the stopping of some of our resupply flights—and the mail.

The Viets won a great battle near here last week. The Cong, two battalions strong, attacked two outposts. One outpost was hit by a diversionary force, the other by the main attack. Viet artillery tore the enemy up one side and down the other, and Viet infantry counterattacked with fixed bayonets. The Cong left behind 110 dead, over 14 prisoners, including 3 officers, and almost 100 weapons. Many more were hit but escaped. Both VC battalions were thoroughly whipped. Among the dead were the two enemy commanders. Let us hope this great victory is repeated often.

When you consider the enemy only attacks when he feels he can win, you can see there are many favorable indications for our side. I often wonder how much more the VC can take. I can see that the Viets are just beginning to fight. They can slug it out for a long time. The Cong sometimes seems to me as though they are becoming exhausted. *(In the meantime the "elites" running the war in our government were somewhere out to lunch. Oh, what a time to pull out the stops and let the enemy have it. One of several opportunities that we didn't capitalize on. Read their books; they didn't even know about this stuff.[5] We had a chance to choke off the North Vietnamese. If Prouty and Newman and others have got it right in their books, we may have war crimes or malfeasance in office here on the part of some top administration people of that time.)*

Kontum, 16 December 1962

As you can probably tell from what you read and hear, the American newspaper and newscaster people aren't sure what's going on over here. There isn't sufficient interest in this war to generate many stories about it back home. I've seen more interesting things that have not been written than the sum total of everything that's been in print so far *(at least that I knew about)*. The tone of the magazines and newspapers seems more optimistic now than it was before. Only the newspaper people don't seem to have the slightest idea about why or what is happening. I have seen all the newspaper people whose stories you have read, I'm sure. Even Robert P. Martin of *U.S. News and World Report*.

When these people come out here to do a war story, they just don't go anywhere or gather any information that's worth a damn. They talk to a few dignitaries and get a general, specially arranged briefing, with [only] the facts that the government wants them to know. They stick pretty much around the wheels and don't get out where the news can be had. Can't blame them too much in a sense. It gets dangerous, and they don't know what they are looking for anyway. Some of the people don't even leave my forts except to travel from one to the other. This wouldn't be so bad if they talked to some of the right people, but often they prefer a drink in the colonel's quarters to interviewing a battalion team, hot out of battle, that may be stopping by briefly for supplies. Once in a while a more enterprising reporter will get on a helicopter with the Old Man and

---

5. Robert S. McNamara, *In Retrospect: The Tragedy and Lessons of Vietnam* (1995).

take a ride out to one of the outposts. There they see another Western-style fort like mine, but smaller, less comfortable, and probably just as safe. The press people don't know much about this war, so they don't write many good stories about it. I think they might like to know more about the war to write better stories. Lord knows this war could provide some fantastic war stories.

But I think it is partly the governments of South Vietnam and the US that are the reason the press coverage is as poor as it is. The government of South Vietnam is extremely sensitive to criticism, and there is much to be criticized here. Despite the fact that there is even more to be applauded, the government clamps down on news coverage because it can't stand criticism. It's an obsession with the Viet government. Our own government cooperates with this unwritten hush-hush type policy of news suppression because we don't want outsiders to know just how much we are putting into this war. It could be bad for propaganda in one or many places, even at home. A lot of talk about the size of our aid here might enable the enemy to scare the Vietnamese people into believing that we want to take over. On the other hand, it might enrage the American voting public into thinking our government is getting us into a good-sized war without informing the people *(thinking about the 1964 election to come)*. Our allies could think the same thing. At the same time, we are trying to keep the Communists guessing at just how much we are committing ourselves here, and more information might tell them too much. And then there is the block of so-called non-aligned nations who might start screaming about American intervention if more publicity about the war were put out. So it is likely to be some years before we see any comprehensive stories about the Vietnam War being put out. *(Apparently we had to wait until the nineties to get enough of it declassified to piece together more of the story.)*

My own impressions of this war are those of an individual who came over here expecting a very bad situation. My job carries me indirectly into the heart of the war in my area here in the Central Highlands. There have been days when I was involved in this war to the extent that war can be ugly. I would look at my comrades and wonder who gets it next. But these days have been few, and most of the time you can hardly tell you are in a war except for the defenses and the weapons no one ever forgets to carry. Life has a very normal and usually comfortable routine. Hard work, busy days, and much responsibility are a way of life that would happen for me anywhere.

I don't know a great deal about the war in the Mekong Delta area.[6] It is a different and generally more active war. It is being fought in flat country with jungle swamps everywhere. The outcome down there appears very much in doubt. Up north in the extreme northern part of South Vietnam, I am very vague about what is going on. The country is similar to that around us here in the central region—mountainous jungle. Their war is probably very much like ours here, maybe a little more quiet.

I was a bit shocked and disappointed with the way I saw things being run when I came over here. Even at that time a lot of improvements were being made, and I shudder to think of the situation here a couple of years ago. I think perhaps the Cong was weaker as well, or they might have taken all the cookies. When I arrived, both sides were building up. The strength the VC appeared to have around here last summer had us all expecting a ding-dong fight. Our casualties at that time, and the fact that the enemy actually prepared twice to attack my main base here in Kontum, made it look for a while as though the situation was dangerously in balance. But at the same time, last August *(1962)*, the fruit of a year's work by the Viets and ourselves seemed to pay off. Just when it looked like all was lost. The tide sure does seem to have turned, in our region at least, since then. In guerilla warfare the guerillas cannot afford to lose a battle. If they appear to lose a fight, the people may think they are losing and their supporters quit. Therefore the enemy never attacks unless he thinks the odds are so much in his favor he will win.

Well, here in our region two main accomplishments appear evident. First, the enemy is attacking less and less. This could mean we are growing stronger and perhaps he is growing weaker. It also means we are keeping him busier escaping our battalions, which are now chasing him around through these torturous mountain jungles. Secondly, when the enemy does attack these days, he is almost always beaten off. Sometimes our people really give the Cong a beating in the process, too. That fight a month ago involved two well-equipped battalions of North Vietnamese regulars who outnumbered and attacked those two Viet outposts. The Reds had over 4 to 1 superiority, and by all odds they should have won. But the Viets had artillery on one of the outposts. It was a fierce

---

6. See chapter 6, "Genocide by Transfer—in South Vietnam," in L. Fletcher Prouty, *JFK: The CIA, Vietnam and the Plot to Assassinate John F. Kennedy* (1992).

fight. Finally, in a last winning act, the little Viet garrison of one of the outposts fixed bayonets and counterattacked against great odds. Both enemy battalions were thoroughly defeated, thoroughly disorganized; both enemy commanders were killed, over 100 enemy were killed and left behind, countless other dead and wounded were carried off, and about 100 weapons were captured. One year ago the enemy would have overrun both of these outposts. What a turn of events. I also notice that attacks and ambushes the enemy pulls off these days are poorly executed affairs. They seem to fizzle out as half-hearted attempts. These two main reasons then, and others I've written about before, lead me to believe that in this part of the country at least, we are winning now. I begin to doubt that the enemy ever had quite enough strength to win total victory in the first place. The task of catching what's left of the enemy (and that's still plenty) up here is a long, tough job. The biggest task of all, breaking the enemy's will to keep fighting, seems nearer, but still a long, hard task. I feel we are determined to go on indefinitely. I often wonder how the enemy can take much more. Everyone's prayers may hasten the day when the enemy just plain runs out of gas.

16 December 1962

I am having a satisfying though at times frustrating time teaching English to a fine group of Ba Nai [Ba-Na] Montagnard tribe boys. They are really great kids. They are going to missionary school here in Kontum. When they graduate upon completion of high school, they will go back to their villages and attempt to raise the literacy rate and, more important, bring Christianity to the people. Christianity is strong in this region, but it varies from village to village. To a good many people who consider themselves Catholics and indeed are baptized, Christianity is still a vague concept.

They are materially poor, but if anyone was richer with God than these humble and happy folk, I did not notice it before. As far as I can see, we Americans immediately hit it off with these mountain tribes people. They closely resemble the American Indian *(or Native American, in later-twentieth-century terms).* They suffer similarly under the Vietnamese. But the French handled them well, and we are also Westerners and we are received well by them.

I have a platoon (50 men) of Ba Nai tribesmen from the local Civil Guard manning my defenses now. The ARVN troops, the regulars, have now been released for offensive missions. The Civil Guard corresponds to our National Guard, but they are, of course, all on active duty. The fact that the Civil Guard and the Self-Defense Corps are now sufficiently trained to take over these static defense duties in many forts frees the ARVN for the task of rooting around in the jungle hunting the enemy. This, plus the fact that the Viet forces are stronger, better trained, and better equipped, accounts for the fact that it looks like we are winning in this region anyhow.

At any rate, life recently has been more peaceful around here than those apprehensive days last August when I twice expected to have to repel howling mobs of hoodlums from the ramparts. I pressed a fortification building program that has made us a tough nut to crack for any type of enemy units hereabouts these days. Here I must be commander, soldier, manager, logistician, businessman, linguist, air transport expert, you name it, yes, even an accountant. So far I've done it, but it's not offensive soldiering. It's more static defensive soldiering. It's more like being the dictator of a village with a weapon in hand and a sharp eye for enemy marauders.

Kontum, Christmas 1962

Our group of Americans has suddenly knitted together since Thanksgiving time into a close group. Our reputation as a group is spreading far and wide in Vietnam—to the honor of the memory of "Terrible Tony" Tencza, our late chief, but to the credit of our present boss, Lt. Col. Robert O. Sweet, who has turned out to be a fine leader. He has truly improved over the months. His outlook on things has matured considerably. It is much easier to work around here than it used to be.

But now to the business of our Christmas festivities. People back home sent us so many decorations that we turned this place into a Christmas showcase. We had 7 or 8 Christmas trees, and our little satellite outposts even had them. We made a huge Santa Claus suit for big Tom Crawford, a 7 foot, 250 lb. Texan. We accumulated a huge bunch of goodies so that we were able to visit our three favorite places in Kontum: the boys school, the orphanage, and the leprosarium. *(We Yanks had written to our families to help us make a Christmas for the local kids. The response was overwhelming, and we and our beloved families back home gave those precious kids a whale of a Christmas. I dearly hope that they didn't suffer from the Communists because of the interest we showed in them.)* We collected enough goodies amongst ourselves, and the government even gave us some money to spend on these people. We bought cloth for clothes, soap, and live pigs to give away. We practiced up a storm and worked up a truly good group of Christmas carolers to serenade these people. We had a great Christmas! Last Saturday afternoon in a big cavalcade of jeeps, we and Santa drove through Kontum with Santa throwing candy to the kids. We drove up to the orphanage and visited the kids. You should have seen it. We gave all the kids a present and [we gave] the French sister in charge some cloth and soap. We serenaded the kids with Christmas carols and then 40 or 50 of us crazy Americans had a ball romping around with those cute little kids. I chose myself a little baby girl five or six months old and had a ball with her.

The next day we went to the boys school, where I've been teaching English, and we did the same thing. What fun. The boys have one of the most beautiful choirs I've ever heard, and they matched us song for song with some very beautiful singing. We gave Father Hoang a gift and some money equivalent of a month's salary for a semi-skilled worker. He can buy many things for his boys with that. The boys were really enthusiastic, and you should have heard them sing, cheer, and try to converse a little with us in English. I was proud of my boys. They sang so well that many of our Protestant officers and men attended Midnight Mass with us just to hear the boys sing High Mass.

Later the same afternoon we went out to the leprosarium village of Sister Marie Louise, that very lively, absolutely fantastic French nun I mentioned earlier. We brought the leprosarium four pigs, more meat than they get in months. And soap, cloth, toys, and candy. We saw things out there that Westerners have never seen. These people were all Montagnards. We were surprised to discover that they had prepared a native Montagnard show for us. Folk songs for Christmas, dances, playing of native instruments, very wonderful indeed. They also enjoyed our show, which was the first touch of American [culture] in person that they really had outside of a few of us visiting from time to time. Many Americans had sent contributions to Sister Marie Louise, but our [visit] was the first people-to-people contact. When we presented the pigs, they went through some kind of a native ceremonial ritual. The one thing I could understand from it was that the pigs had had it and were not long for this world.

Christmas Eve we threw ourselves a big party and decorated our Christmas trees. All during the days up to that time our Vietnamese friends, soldiers, and employees would drop by and leave us Christmas cards and touching little gifts. My people would sneak into my room and leave a card or a little package of fruit or a bouquet of flowers. We were all really touched by these things. Two guys bought their maid a bicycle for about 1500 piastres. She cried and cried, she was so overwhelmed. In her life she probably couldn't afford one. I found our guys sneaking out and taking toys to the kids of the people who work for us. It has all been quite something. We gave a feast for our guard platoon one morning, roasted a big pig. One of the 22nd ARVN Division officers shot a huge deer and presented it to us as a Christmas gift. *(How could our American national leaders insert us into a situation like that and then years later abandon these people to a horrible fate and also abandon those of us who were the soldiers they put there? For each of us our own private sorrows.)*

Christmas Eve we piled into jeeps and headed into Kontum to go to Christmas Midnight Mass. *(Kontum was a big Catholic town and had a French Catholic bishop.)* What a sight awaited us! The many Catholic churches of Kontum were beautifully decorated and lighted. Most of them had altars erected on the front steps outside of the church so that the Masses were actually outside. The air was brisk and a little chilly, but the night was beautiful. All of the parishes of Kontum got together in a Mass procession through the heart of town and then fanned out to their respective churches for Mass. Everything was gaily decorated. Each parish had a beautiful float mounted on a truck. All the people marched behind carrying candlelit lanterns. What a procession of lights! At the rear of each parish was the priest in robes and the altar boys. We were astonished at the many thousands of people we saw.

When the procession finally ended, we drove out to the high seminary on the outskirts and saw how beautifully they had decorated their buildings of old French architecture. A colorful procession of seminarians was making its way across the grounds to the chapel chanting prayers. We dashed from there to the shrine of Mary, which is at one end of a street. At the other end two blocks away is the rather modest cathedral of Kontum. Each sight grew more spectacular. The shrine was beautifully lit up and decorated, and people were praying everywhere. You could look down two blocks and see the cathedral also magnificently decorated and lit up. A beautiful sight in the dark night. And lining both sides of the street were handmade lanterns on poles. Made of tinted paper and lit by candles, they were a fabulous sight. We had Neil Sheehan, a newspaper correspondent for UPI [United Press International], with us and he did a story on it.[7]

We drove down this fabulous street and passed the cathedral to head for the boys school nearby. All the buildings were lit up at the boys school in the same manner. Most of the lights were all torches and candles everywhere we went. But they seemed so brilliant and colorful in the clear, dark night with their colored paper and cellophane covers. We were nearly late for Mass, but we made it and it was as beautiful as we expected. Our dear friend Father Hoang sang a beautiful Mass, my boys sang out well, and my many Protestant American friends were impressed. I never saw so many Protestants in a Catholic church. The humble but beautiful faith of the mountain folk seems to impress them.

Two of my guys are former ministers, and we have a couple of American Protestant missionaries in town. *(These were the Smiths, from the Christian and Missionary Alliance.)* The Protestants really get out to church here also. They are devout, open-minded church-going Protestants. Many of us are just getting back from Mass on Sundays when they are leaving for services at the Smiths'. They have their own services when the Smiths are up in the hills. There is some joking, a few discussions, but every consideration and courtesy is shown for all religions. We must, after all, get along with those of our Vietnamese counterparts who are Buddhists. But here in Kontum, Buddhists are few and Catholics overwhelming.

I practiced my Catholic faith devoutly, not for fear of death in combat but because of an earlier commitment. As I mentioned before, my friend Lt. Humbert Versace was ridiculed by some of his West Point buddies for having a heart after God, so I took him under my wing, and they didn't mess with him when I was around. They didn't mess with me either because of my seeking after God. Unfortunately, Versace was caught and executed in 1965, but I know my friend died a Christian.

General Timmes, who was second in command (Chief of MAAG) for a long time in Vietnam, was also a spiritual man. He would come up to Kontum to visit us, and the first thing he would do was ask for Captain

---

7. *Webster's New World Dictionary of the Vietnam War* (1999), 366.

Frankwicz so we could go to Mass together. We would go there, two like-minded soldiers seeking after God, and it would seem to help us. On the way back I got to have a lot of private talks with him of the "networking variety," far beyond the realm of what would have been normal. Unfortunately, this kind of networking didn't seem to help the war effort, as the reader can see from the rest of this story.

Kontum, 26 December 1962

Today Francis Cardinal Spellman of New York came all the way out here to visit us. What a saint the great man is. Some weeks ago it was rumored that Cardinal Spellman would visit American soldiers in the Far East on his way home from the great church council in Rome. He has always been a great one for visiting American soldiers away from home at Christmastime. Some of my men got together and wrote him a letter asking him to come to Kontum at Christmas and bless the monument that Ed Mixon and the outfit and I had built to Col. Tencza's memory. Doggone if he didn't come! Following along behind him were a host of church dignitaries and generals, etc. But the cardinal came out to Kontum to pray over our monument and spend some time with us. He looked old, weak, and failing but very happy, very chipper, very brave, with a wonderful sense of humor. If ever I met a great saint, there was one. He was like Father Dismas Clark (whom I had met as a young altar boy),[8] having a strange and holy glow all about him. *(I guess it was the Holy Ghost.)* He chatted with us and blessed our monument. Reporters and cameramen fell all over each other and us. Maybe you'll see it on TV. He said he knew Janesville, Wisconsin *(my hometown at the time)*, and was interested in what we were doing. He softened up the toughest soldier with that smile and way of his. All of our religious friends thronged over to see him and were very pleased with the way he gave particular honor to Sister Marie Louise and Doc Smith. What a lineup of saints! Incidentally, I gave the baby clothes [likely referring to clothes his wife Peggy had sent] to Sister Marie Louise. Any more you can get will be helpful. Do you know Cardinal Spellman gave us a check for $ 500.00 to buy some things for our local charities? That is the equivalent of

---

8. Father Dismas Clark, S.J., ran a refuge house for men in St. Louis during the middle of this century. He conducted a men's retreat sometime in the late forties at St. Patrick's in Janesville, Wisconsin, that was a very spiritual meeting. I was an altar boy for him.

38,000.00 piastres. What we can't do for our friends with that. I guess he figured when it comes to charity nobody could do more than GI Joe. We should be able to keep full tummies for our kids for a long time.

27 December 1962

(My mother sent me a Bowie knife for Christmas, and it was about to be a very useful gift, as I was shortly to be assigned to the Civil Guard as an advisor.) The Bowie knife is the envy of all Montagnards, Viets, Americans, and Frenchmen around. I keep it safely locked up when not wearing it. In this country, that knife is regarded as a Rolls-Royce or Mercedes-Benz auto is back home. The goatskin gloves are excellent protection against the sharp and thorny vegetation we have to beat back every couple of weeks. The machete is useful for hacking at the stuff. The old skinning knife finds its principal use in ripping open supplies. We come by an occasional pig or deer, but my Viet soldiers do the skinning and cleaning.

I never cease to get the impression that with the exception of a goodly number of French and American innovations, we are living in the world of the last century here. The power tool and machine is a luxury. Only the most powerful men in the province can command the use of such tools. I am fortunate enough to be able to borrow such equipment from time to time. But my men and I do most of our work by hand with the crudest tools. Sometimes we make one type of tool do the work of several types. You should have seen us engineer and build the system of roads, culverts, and ditches we have here.

The Bowie is strong, sharp, and beautiful. I demonstrated its qualities to a squad of my Montagnard-Viet security troops, and their eyes bugged out an inch. First, I took a thread and dropped it on the blade. The thread fell away in two pieces. (I saw Saladin the Great do that in a movie once to impress Richard the Lionhearted.) Then I selected a sizeable branch on a tree and figured to do well to knock it off with three well-placed strokes. Off it came with one deft slice. That really brought the oohs and aahs. (I saw Richard the Lionhearted do that one in front of Saladin in the same movie.) To us and the Viets, that just proved I got a swell knife from my mother. But to my Montagnard troops, who I really depend upon for work and protection around here, it worked a calculated effect upon their primitive minds. I am their leader, and as the leader, mine must be the sharpest, strongest knife. The knife and the man, they figure, are one. I always wanted to pull off a trick like that after I saw that Cecil B. DeMille movie. *(They started calling me Dai Wee Con Yow Yiee [Đại Uý Con Dao Vĩ], meaning Captain Big Knife. What a show-off! But stuff like that builds a leadership image.)*

In addition to the large Bowie knife that I carried instead of a machete, my personal weapons at this time included a Thompson submachine gun a la World War II, which I had to unpack and clean off the cosmoline from. It may have been packed up somewhere since then. I also had a .38 Special, which for some reason I carried more often than the Colt .45 pistol. I also had two hand grenades, which I carried attached to my webbing field gear. I had a small .25 Spanish automatic that I borrowed from my brother-in-law Al and which I carried concealed. Later I was issued an M3 "grease gun," the World War II replacement for the Thompson, which hung around for a lot of years in the army.

I had become an expert in all infantry battle group weapons on the way to winning the Expert Infantry Badge back in the 19th US Infantry in Augsburg, Germany. When some of us went to the rifle range to practice with our weapons, I got so I could fire the M3 single shots and consistently hit the bullseye at one hundred meters. They told me that was pretty good, since the weapon was designed to be fired full automatic at fifty meters or less.

(Here I recap the Christmas festivities in the letter to my mother.) One guy fixed up a Christmas tree by taking a grisly looking piece of a dead tree and for ornaments [using] bullets, grenades, mortar shells. We got a good laugh at ourselves. We went to great pains to get us a Santa suit and we finally had one tailored in Kontum. Santa was a huge, 7ft, 250-pound Texan by the name of Tom Crawford. A captain and an artillery man by trade, many a Cong has fallen under his Viets' guns. A couple of days before Christmas we formed a big jeep cavalcade and roared through Kontum with Santa in the lead throwing candy to the kids. We went to the orphanage run by a French Sister Anne, and what a group of excited little ones we found. Santa was so huge that some were scared of him at first, but after we sang a few Christmas carols, all was well. We distributed soap, cloth, toys, and candy and had a ball running around the picturesque grounds of the place with the kids for a few hours. We stayed much longer than we planned.

The next day we went to my Montagnard boys school and the fabulous boys choir and us sang carols back and forth to each other. Santa distributed more goodies and once again we stayed late. Lastly, we went to the leper village run by the famous Sister Marie Louise. What a surprise. They arranged a real show for us, with the little kids doing dancing and singing. Then the grownups, among whom there were some really bad cases, got together their village band and gave us a blast of tribal music on a weird assortment of instruments. We gave our program for them, and Santa distributed pigs, cloth, soap, and our toys and other gifts. We all gave money too to these places. So you can see that you all gave us a very fine Christmas away from home, and you helped us pass a little on to some folks who never had one before. For once the ugly American wasn't so ugly.

In discussing this Christmas among ourselves, we all agree we really missed our families. But we also agree we were not sorry to be here. Out in the jungle the enemy grumbled and growled and belittled us and boasted how many of us they have killed. (They claimed enough to have gotten all of us three times over.) The enemy lost a great battle, and the strangest, holiest thing about it was that none of us thought of it that way until Christmas was over and the rumors of the Cong's reaction filtered in. Perhaps there is more of the spiritual dimension in this world than we realize.

At this point, December 31, 1962, the record says there were 7,900 army personnel in Vietnam.[9] Again, comparatively few were fighting types.

New Year's Day 1963

Today I had to fly by helicopter to Khon Brai [Kon Braih]. Spent part of a very interesting afternoon out there. To enter the place you have to go through a maze of barriers. There are pitfalls and traps, everywhere. Mines, sharpened bamboo spikes, and barbed wire covered by mortars and machine guns surround the outside. There are several layers of these all well covered by weapons. The last layer next to the wall is really rough. After all that you come to a moat. The inside is row after row of needle-sharp stakes. Once you are inside the main defenses, you are still impressed by the fact that the Viets have apparently not overlooked

---

9. *Webster's New World Dictionary of the Vietnam War*, 484.

anything. Every open area inside the main wall is also covered by weapons in case the enemy gets in. Trenches connect one bunker with another. There are 2–3 companies there as a garrison. One company is constantly patrolling in the high surrounding jungle-clad mountains. The place reminds you of a cross between an old Western fort and a medieval castle. It would take quite a battle for the enemy to take the place now. *(They took it from the French and had attacked it once or twice by this time under the Viets.)*

Last night at midnight (New Year's) they fired all the guns in celebration. Must have been quite a sight. Our advisors up there are Captain Parkhurst and Sgts. Stude and Chadwick. The battalion commander of ARVN 2nd Battalion 41st Infantry commands the fort and the troops in the area. He is my good friend Captain Binh. If ever anyone was a tough-looking character, Captain Binh is. He lost one eye in combat some time ago, and the results have made him look fierce and rugged. Actually he has a very pleasant personality, and I really enjoy his company on rare occasions when we get together. He knows his stuff, is a good fighter, and doesn't need much advice. In his case and others like him, where the Viet commander is actually more experienced than his American advisor and well able to do the job, the value of the Americans is used differently. Instead of advising the commander, the Americans will spend most of their time helping him. Helping him train his troops, helping him administratively, and helping him solve the host of problems that plague a unit in combat. The advisors form a sort of staff for such a commander. This makes a good combination, because most Vietnamese staffs aren't up to par yet. The hope is they will learn from our people. But it will be a long, hard road yet, because many of their people just have not got the stuff or the drive.

Upon finishing at Khon Brai we whisked back to Kontum and I changed from battle gear into TWs [a tropical-weight uniform] and attended the colonel's New Year's Day reception. This, complete with a punch concoction we whipped up, was a chance for us to behave like humans again. We all dressed up fit to kill just to please ourselves and to remember how to do such things. After the "reception," I and my Catholic buddies made the evening Mass at the 22nd Division chapel.

Toward the end of my tour of duty as base camp commander, I was assigned a short, heavy-set black first lieutenant by the name of John Lewis, who hailed from Washington, DC. Very soon, John became worth his weight in gold as my executive officer. He was competent, diligent, and had a "can do" attitude. We were in fact becoming good friends as well.[10]

One day John took the supply run up to Tan Canh along with one of the sergeants. In the early evening, someone came in the CP [command post] and told me, "Captain, we've got a couple of wounded out here." Out behind the CP was a jeep with two unconscious Americans in it. One was the sergeant, the other was unrecognizable. I jumped up in the jeep and looked for the name tag on his field jacket. It said Lewis. I blurted out, "Oh, no! John!" It was John Lewis, so swollen up we couldn't recognize him. He opened his eyes and mumbled, "I'm sorry to let you down, Cap'n Frank." Then he passed out. Tears welled up in my eyes as I choked out, "My God, John, you get wounded in action and you apologize." Their jeep had hit a mine on Route 14 and flipped over on top of them. I knew several American heroes, and one of them was John Lewis. Each of those losses hurt, and John Lewis's loss was felt deeply by me personally.

---

10. I came in when the army was integrating the races, and I thought it was going well enough to be considered a non-issue. I had a number of black officers for close buddies, and about six of the ten best officers I knew were black. They were a credit to their people, a credit to the service, and my good friends.

CHAPTER FIVE

A NEW ASSIGNMENT

JANUARY–APRIL 1963

Kontum, 5 January 1963

WHOOPIE! I take over the Kontum Province Civil Guard Battalion and the SDC (Self-Defense Corps) Training Center advisory slots. A new captain, Perry D. Marks, a slightly older fellow, takes over for me. The local civil guard has responsibility for defense for Kontum Province (much as the Wisconsin or Missouri National Guard do back home).[1] The troops man outposts and form relieving forces for outposts and villages under attack. Occasionally they also take part in pursuit operations. The troops of the battalion are scattered all over the province of Kontum, a platoon here, a platoon there, etc. There is even one platoon guarding my MAAG compound, so I will be no great stranger to them. As I see it now, I will be working out of Kontum mostly, with frequent visits to various outposts. The other job I have is advising the commander of the Self-Defense Corps Training Center in Kontum. This center trains (or

---

1. For a good narrative of the background of the Civil Guard and the Self-Defense Corps, see Robert Shaplen, *The Lost Revolution: The U.S. in Vietnam, 1946–1966* (1965), 137ff. Shaplen, a University of Wisconsin man, wrote an intensive narrative of events in Vietnam from 1946 to 1965.

attempts to) groups of villagers to defend their villages against the Cong. This also should be interesting.

The commander of the battalion is Captain Bui Thi Minh, and his exec is Lt. Thich. Minh is a tough old combat veteran of 44 who formerly commanded a parachute battalion in the Delta area. I doubt I can give this fine-looking soldier much advice, but I do know I can help him some. He'll probably do most of the teaching, which is OK too.

The base camp commander job I had been doing up to that point was really more prestigious than I had thought it was at the time. I wanted a combat battalion, like the Vietnamese Airborne Brigade battalion I'd expected at the beginning. I felt like I had "been had" by my West Point friends, especially since I felt like I was as good a professional soldier and with as good long-range career potential as any of them.

Furthermore, I was one of only three who came out of the army's language school at the Presidio of Monterey really, and I mean really, speaking Vietnamese (albeit not expertly). Those language skills were supposed to be used advising the combat effort, not doing business in the marketplace as base camp commander. "Stupidus maximus!"

I finally got reassigned to the 10th Civil Guard of Kontum Province as a senior advisor. My counterpart was the very professional Captain Minh, who had been fourteen years at war against the Communists. He reported to the province chief (similar to being governor of an area the size of New Jersey), who reported to the president of South Vietnam, Ngo Dinh Diem [Ngô Đình Diệm].

I was in my middle twenties, the guy put on the spot, a captain in the paratroopers, an elite infantry expert in our army, able to use his Vietnamese language, but I felt awfully humble next to this patriotic soldier [meaning Minh]. It was an awesome command—three thousand men, twenty-six rifle companies, a brigadier general's command in our army—but he only had three competent Vietnamese officers, including himself, to do the job of hundreds. I told him, "I have no combat experience, but I have led some of our best American combat units. I am well trained. I will help you wherever you say; I am yours to command." We liked and respected each other, so I found myself often with the maneuver element, a battalion-sized unit counterattacking enemy main force battalions.

There should have been over three hundred officers in that unit, but there were probably less than forty. Of those, only four were competent by our

standards: Captain Minh, his exec Lt. Thich, and two company command-
ers. Each company had a number of competent NCOs who really ran the
units. Compare that with my experience in our units, say when I was in
Germany, and it looked like this: A several-hundred-man task force under
a battle group would have maybe a dozen US combat arms officers. Proba-
bly ten would be pretty good officers. There would probably be forty or fifty
competent NCOs. In addition, in those days we had a 50/50 draft versus
volunteer army, and with a large number of well-educated and experienced
draftees. Among the soldiers, then, there was much leadership material. A
typical US rifle company could suffer heavy combat losses and still come up
with leaders from the ranks to carry on. Our World War II and Korean
War experience proved it. The Vietnamese, however, couldn't find a leader
in a rifle company after the key one or two people went down, and the situ-
ation often disintegrated. It was a cultural issue, and I didn't think it was
appreciated by some of our key people. The further that unwise leaders got
from the fighting troops, the more they became enamored with maps and
stats. A visit now and then allows them to convince themselves they are "in
touch," when they couldn't be farther from it.

12 January 1963

I have finally succeeded in getting a new job. [On] the 19th I begin
as advisor to the commanding officer of the 10th Civil Guard here in
Kontum. The new job will be more down my alley. This is a monster of
a unit for a battalion. It has 11 companies. Battalions in our army have
3 maneuver companies. It is more the size of an old American regiment
or brigade (which contained *3 battalions*!). The mission of the 10th is
the security of Kontum Province (about the size of New Jersey). Most of
the troops are scattered far and wide manning the many outposts in this
province. One rifle company is protecting the airfield at Pleiku. But some
units are a mobile reserve. There is even an armored-car outfit. Some
units are also in training. It is the purpose of this command to take over
the provincial security so that the 22nd Division ARVN troops are freed
for offensive actions.

(The entire Civil Guard command, including the SDC, was about three
thousand men in twenty-six companies, the size of a typical brigade com-
manded by a brigadier general. At this point in the war, Vietnamese units
were commanded by officers a couple of grades in rank lower than the world

standard. Thus, the ARVN 22nd Division was commanded by a full colonel, whereas typical battalions are commanded by captains. For that reason, we supplied advisors of equivalent rank. Captain Minh commanded a brigade-sized unit, which in other nations would be a brigadier general's job, and it was my privilege to help him. It is interesting that this worked out just the opposite as earlier in American history (during the American Revolution and even in the Civil War), where you had units called regiments commanded by full colonels that were actually about the size of modern companies, which are commanded by captains and often lieutenants. Captains in those days commanded companies the size of today's platoons, fifty men or so. But in the mid-to-late twentieth century, the captain commanded larger units and with far more complex weapons systems, requiring considerable education.)

One of the outposts, Plie Mrong, was in the news recently. That was a particularly vicious battle that we lost. We had a bad week that week. The enemy wounded two Americans, killed 29 Viets, captured 100 weapons including 50 submachine guns, and overran Plie Mrong. There was considerable excitement that night. We discovered the enemy unit making the attack was commanded by an East German captain! The enemy losses must have been heavy too, because the garrison put up one hell of a fight, and only one portion of the fort was actually overrun. Our artillery rushed down the Pleiku road (Route Nationale 14), unlimbered, and pounded the hell out of the area, which must have hurt somebody, and the next day we flew in two companies and almost netted them. But the wily so-and-sos gave us the slip again. That was the best-executed enemy attack in this region that I can remember. That particular enemy unit will probably give me plenty of headaches. The fifty automatic weapons they captured is what makes my stomach tight.

Tell them [the folks back home] that I'm proud of the Badgers and the Packers.

I am fortunate that Captain Minh is a man whom I like and respect. It should be a pleasant association. He also speaks good English, so that I will not have to strain too much with my Vietnamese. Somedays I get tired, and talking to these people seems to take so much out of me. I need a rest to rebuild the energy I used to have.

More about the Plie Mrong incident: Early in 1963, in typical actions of that period, an enemy main force battalion (six hundred to eight hundred

men) would come across the Cambodian–Laotian border and strike vil-
lages in the mountain highlands in Kontum Province. These villages would
be defended by our Self-Defense Corps soldiers, who lived in them. The head
man in the village had been issued a shoebox-sized radio to call for help.
The villages and the forts had large arrows mounted on a post that could be
rotated to point toward an attacking enemy. At night the arrow could be
touched off on fire to direct friendly aircraft.[2]

These operations happened weekly, as I recall. On this particular occasion, an enemy main force battalion struck the village of Dak Sut quite a ways up north on Route 14 from Kontum. By way of interest, rumor had it the enemy battalion was advised by an East German captain. They hit Dak Sut at night and a fierce struggle ensued. The village headman called in for help on his radio. We couldn't get helicopter lift from Pleiku for our counterattack strike force, so we requisitioned trucks and Lt. Thich and I led a company-sized strike force up Route 14 to the relief. The enemy apparently put an ambush force on the last mountain before Dak Sut. They mined the road at a narrow spot with a dropoff on the other side. Dak Sut was visible from that location in the next valley. Another party consisting of two duce-and-a-half trucks with ARVN troops on board was heading north in front of us and apparently ran into the ambush. The mines blew up the trucks, and wounded were lying about as we came on the scene. I jumped out of my jeep and signaled back to our trucks for our Montagnard troops to immediately fix bayonets and charge up the mountain in a surprise bayonet charge. Heavy enemy covering fire over their ambush never materialized, and the mountainside went quiet other than the shouts of our Montagnard soldiers scrambling up the mountain with fixed bayonets.

---

2. Susan Katz Keating, "Plausible Denial," *Air & Space* (May 1997), 47. "'We worked out a system with [the ARVN] at these little outposts, where they would set up a flaming-pot system pointing out the direction of the enemy,' [says US Air Force Colonel Benjamin King]. 'Later on it became a flaming arrow.' The large arrows, covered with woven bamboo, were laid directly on the ground. 'They would point the arrow in a certain direction, and it would come over the radio: "Drop your ordnance 200 meters away from the fire arrow," or 100 meters, or some such,' says Farmgate [King's unit] pilot Joe Kittinger. 'Sometimes it worked very well,' says King. 'Other times it didn't work worth a damn.' When it didn't work, the fire arrows became merely another part of the confusion."

Lt. Thich thought I overreacted. I learned from my readings that the *only* cases in history where ambushes were successfully busted by counter-guerilla forces were cases where there was an immediate violent counterattack, the last thing the enemy expected. It works because it is the unexpected; it's a counter-surprise. If your troops are regulars and theirs are not, they can't stand up to you, and they know it and they panic. You get seconds to do the right thing. It's like an athletic event. If you do it every time, when it counts you'll do it right. Otherwise risk going home in a body bag.

I thought the enemy was there and had withdrawn quietly and in some distress before us. I was curious at Lt. Thich's response to the situation. In any event, if Thich was right or not, I did the right thing, and there was no bloody fight at that spot. I then turned my binoculars on the village of Dak Sut, down in the valley several kilometers away. Columns of smoke rose into the sky from the scene as a result of the fight, and the enemy main force battalion had withdrawn, as the pattern of the time was that they did not wish to engage the relief but instead withdrew to the sanctuary over the Cambodian–Laotian border to refit for next week's episode. Our hope was to get them pinned in a fight and then bring in the resources to destroy them. *(We were doomed from the start on that one. We had to go through top-heavy bureaucracy riddled with enemy agents to get anything to happen. It seemed in those early days that if we could just get those people out of our face, we could achieve amazing results, if for no other reason than on the basis of surprise, for the enemy seemed to expect us to perform no better than our inept high commanders.)*

We had about five companies of Montagnard soldiers that were in various states of basic training. Just outside of Kontum City we had a rifle range where these troops would go for firing practice. I thought it odd that they would have to march out to the range with loaded weapons ready to fight even before we trained them.

Kontum, January 1963

Weather here continues dry, dusty, and windy, with comfortable temperatures. Sometimes the nights get downright cold. At least they seem so, because we don't have the clothes for it. The roads, which can get axle deep in the mud during the monsoon season, are now covered with several inches of fine, powdery dust. The stuff gets into everything when you travel. Every vehicle behind the first one is enveloped in huge billowing clouds of dust that gets into your lungs, weapons, everywhere. The hills and mountains here are very steep and rugged. The jungle claims every-thing. It is so thick you can only see into it for a few yards. Any move-ment off roads and trails requires that you cut your way through with machetes step by step. Anything on the map that is off the roads or out of forts and towns is no-man's-land.

When the VC hit, they come from the mountains and also from the Laotian and Cambodian borders. This vast area covers thousands of square kilometers. From Kontum to Dak To [Đắk Tô] is probably over 100 km (about 70 miles). From Kontum to Pleiku is 45 kilos (about 30 miles). Rapid travel is only possible by helicopter. *(Civilization sticks close to the road.)* Jeeps and armed convoys ply the roads through the muck and mire of the rainy season and the clouds of dust of the current dry season.

Kontum Province has a population of over 100,000 people, mostly Montagnards located in isolated villages in the hills. The many forts you see are actually old-time Western or medieval-style affairs manned by 50–200 or more men. Most of these troops come from the twenty-six companies of my Civil Guard, SDC. The 22d Division ARVN also has troops at these outposts as we previously mentioned. ARVN uses the outposts as bases to go out and beat around in the bush looking for VC. They don't do that as often or as effectively as we would like. Most of the forts are good protection from small arms fire, but they would take a real beating if the Cong ever brought up mortars. Khon Brai [Kon Braih] did fall, in fact, in 1961 under enemy mortar fire. That was the story we read about in the *Saturday Evening Post*. The place has been rebuilt much stronger since.

Some of these forts are real pest holes. You wonder how the troops can stand them. Others like Khon Brai and Plei Kling [Polei Kleng] are downright comfortable. Most of these forts are located to protect key towns or bridge sites, or some key portion of a road. You really get the impression of being in a vast, empty loneliness at the end of nowhere at most of these places.

One afternoon we were heading back south to Kontum after checking something up north. We'd just crossed a bridge where we had a blockhouse (a small fort protecting the bridge) when, "Whoomph," explosion just out of sight over in the river. I turned and looked at the sergeant and he said,

"Some of the men are fishing." I said, "Loudest bait I ever heard!" He said, "Come see." So we stopped the jeep and made our way through the jungle to the riverside. There were a couple of the blockhouse garrison in the water gathering stunned fish. Want fish for dinner? Throw a hand grenade in the river!

On another afternoon, we were having a big command confab with all the ARVN and Civil Guard middle-management wheels and advisors up on a hill overlooking a river up north on Route 14. When the meeting ended, I jumped in my jeep to head for my troops, and as I pulled out, one of the hand grenades hooked into my web gear suddenly popped off, made a graceful arc through the air, and landed at the feet of the august assemblage. I slammed on the brakes, jumped out, and retrieved the grenade, all in a flash, and drove off feeling like a fool. Not a one of those guys flinched, not a word was spoken. I was totally embarrassed, but at least I saw strong evidence that I was working with some brave and dignified men.

In theory, when the enemy attacks a village, the nearest forts send troops. More troops are brought up from Tan Canh [Tân Cảnh] and Kontum. In fact, the enemy usually escapes before the trap slams shut. Our most effective single weapon in this province has been the artillery. We've got it spread around pretty well. What isn't already out can hustle up and down the roads and start firing from anywhere. And these guys are real good artillerymen.

We had an attack this week, which was a very bold enemy action. Three days ago the enemy swept out of the mountains between Dak Nakram and Dak Sut and attacked two Montagnard villages. They easily overran the villages, taking several prisoners. Two relief columns set out, one from our 104th Combat Company at Dak Sut and the other from 897th Combat Company at Dak Nakram. The fight lasted all day long. Our troops were held off by a river they couldn't cross. Our troops did not do well. The one bridge over was well covered by a small enemy force. We lost all our officers, an NCO, and six men in an unsuccessful attempt to storm the bridge. After that, nobody could get the friendlies moving. Worse, this left 104th Combat Company without any officers at all!

While our men were pinned down in front of the enemy, 2nd Battalion 40th Infantry ARVN found a fording spot downstream and started to sneak around the enemy. At the same time, guns of the 22nd Division Artillery were wheeling into position near Dak Sut. But the wily enemy pulled out to avoid the trap. It appeared a VC victory, but then the artillery opened up and apparently they caught the enemy a good one. The battalion commander and I arrived from Kontum when most of the hullabaloo was over. We got a message from the enemy commander demanding we stop that blankety-blank artillery or he would execute two prisoners! Unfortunately, we got the message too late. We got another message before we could act, stating that they had already shot the two prisoners. I doubt we could have been able to stop the fire anyhow. An American would have thought it over, but these Viets would have kept right on shooting.

The thing that disturbed me was that the Viet commanders were content with merely driving the enemy off. They took no *(further)* action to pursue or attempt to trap the enemy. They settled for the status quo. I imagine the two villages lost a couple killed and 25 wounded or so. My troops lost 10 wounded. I don't think the ARVN lost any. The enemy probably lost more than we did. Neither side gained a damn thing. All in all, the enemy can afford to lose people much less than we, and he was probably left weaker after the battle. But as long as no one goes in and ferrets him out, he can continue to make a nuisance of himself.

I do not know what we can do to make these Viets go after these guys. I could see in my mind's eye the way American troops would have gone in. If we hadn't smashed that enemy force on the spot, we'd be after them until we did one way or another. Even if we didn't have any more to fight with than the Viets. All the enemy had to do to get out of it was cross one mountain ridgeline and hide from the artillery. It is difficult to move around in this country and know that the enemy is beyond the next ridgeline getting ready for his next effort with relative impunity. We have superiority enough to really have a showdown, and for the last five months we had the Cong down in this region. But we couldn't get the Viet commanders to go in and deliver the coup de grace.

These Montagnard and Viet soldiers are good guys. Because there are not enough steel helmets to go around, they wear their navy-blue berets. They won't even take those berets off when they are wounded, they are so proud of those little hats. I saw one Montagnard soldier being brought in wounded on a stretcher. The guy was holding his blue beret on his head for dear life and he was grinning just as proud of himself as he could be. We were proud of him too.

The battalion commander, Captain Minh, and I visited the wounded in the battalion dispensary this morning. They all tried to sit up at attention when we walked in. We went to each man. Since this is the Vietnamese New Year, Tet [Tết], Minh and I gave each wounded soldier a gift. In the dispensary each soldier had his blue beret hanging at the head of his bed. Those guys really had their eyes light up when we gave them more chow than they've seen in a long time. All the battalion officers present participated. We stopped in front of each bed and the NCOIC [non-commissioned officer in charge] of the dispensary would rattle off in Vietnamese how each man had been wounded. Were they proud. I really like these Montagnard and Viet soldiers. They are the guys bearing the brunt of this war, and their entire country depends on them.

While American medical support took a big step forward in the Vietnam War, largely due to helicopter evacuation, South Vietnamese medical operations were largely inadequate, although subsidized and overseen by the United States. The enemy had crude medical facilities. They faced health problems due to malnutrition and were forced to forage.[3] We Americans were larger, stronger, healthier than the Vietnamese of both sides. The enemy of the period, often on the run, often suffered malnourishment. One would think that later, when well-fed US fighting units entered the mainstream of the fighting, somehow our better health and strength would show through. I have not seen any evidence of that. This no doubt had to do with the fact that the enemy held the initiative.

25 January 1963

I have little time to influence this unit and they really need help. The line-and-file soldier is good, but the caliber of officers and NCOs is bad. Besides, they don't have nearly as many officers and NCOs as they should have. Nor do they have the equipment they should have, but this is generally a result of ineffective leadership.

Our troops are stretched far and wide. We are augmented at many locations by the ARVN units of the 22nd Infantry Division, which use our forts as operational bases. The present concept is to get more and more Civil Guard troops into static security missions such as outposts, towns, and roads, and free ARVN for offensive strike force missions.

---

3. *Webster's New World Dictionary of the Vietnam War* (1999), 248–52.

Progress has been made, but the 10th Civil Guard is not ready yet to defend the province on its own and probably won't be until the end of '63 at least. *(DREAMER!)*

The Self-Defense Corps (SDC) is made up of local villagers who are trained and armed to resist VC attacks when they come. They perform their normal civilian functions and take up arms for periodic training and when attacked. The enemy almost always preys upon the SDC, as they normally go after small villages in their attacks.

Civil Guard units, which are the same as our National Guard on active duty, usually respond to enemy attacks and get involved in plenty of fighting also. Usually the Cong do not try to fight ARVN regular units around Kontum, although they have managed to successfully down south *(in the Delta)* recently.

Now the enemy is getting pesky again. They seem to be able to roam about at will. They don't seem to have any really decisive punch, but they ought to be thrashed before they build up anymore. When the French fought their last famous campaign in this region, in '54, they were outnumbered. The Communists seized outpost after outpost on both main roads south to Kontum—the same outposts my squads, platoons, and companies are now occupying. Then they side-stepped Kontum and captured a couple of outposts south on the way to Pleiku. That did it. The French withdrew to Pleiku. The situation was different in those days. The Reds were far stronger and the French numerically fewer than present friendly forces. Also, the French were less popular than the national government of Vietnam. Most of the outposts are probably stronger now too.

29 January 1963

Back to the north again today. Visited outposts of Dien Bien [Điện Biên], Dak To [Đắk Tô], Po Ko Ha [Pô Kô], and Dak Nakram. Ate a rather delicious lunch in the 95th Combat Company CP [command post] at Dak To. Hope I don't get the runs from it. The last time I ate in the field with these guys I got the shakes for a couple of days. Maybe I'll get used to it soon.

A large body of VC was detected coming in from Laos yesterday heading for Po Ko Ha. I think they might be Pathet Lao turned VC for the sake of engaging us. After they were discovered, they seem to be going back to Laos again—maybe to try again later. The battalion commander and I went up north to send out patrols, ambushes, and spies and runners to screen the area. With the few people we have for such a vast area, we would have to have some luck to spot them. But the enemy force is large, so it will be harder for them to avoid us. Anyway, it looks like they are backing out. This little game of cat and mouse goes on all the time. North Vietnam sends troops and subversives everywhere. What you call them depends on which country they happen to be in at the moment. While in Laos, they are known as Pathet Lao. When they are in Cambodia, they are either just bandits or not acknowledged to exist by Cambodian authorities. Here in Vietnam, after they sneak in here from the north or one of the other two countries, they become known as Viet Cong or Viet Minh. It's pretty difficult to prevent their sneaking in when the borders themselves are hard to find. Oftentimes our people have found themselves accidentally on Cambodian and Laotian soil and have had to sneak back out again before discovery. The Laotian and Cambodian governments don't give a damn about the Reds using their territory as a base against South Vietnam, but they do raise hell if they catch Viets or Americans on their soil. *(That should have telegraphed to us who the neutral parties thought the winner was going to be!)* Actually, neither government could do a lot about it one way or the other, since apparently neither government has control of their frontiers either.

South Vietnam has limited capabilities of carrying on operations against Reds in Cambodia or Laos, but this would involve them in war with these two countries. They have all they can handle right now. Probably the US should play the role of the powerful ally and put pressure on Laos and Cambodia to maintain genuine neutrality. But our foreign policy in Southeast Asia is such a fantastic mixed-up mess that we are trapped into watching things go on as they are. We are giving support to a cranky Cambodian government whose troops have fought skirmishes against our allies, the Thai and the Viets, and also, on rare occasions, the Communists. This same government is allowing the Reds to move around and use their frontier as a base against us. And we *(that is, President Kennedy and his small clique of "elite" advisors)* actually set up and strongly supported the Laotian situation, and it is from there that the majority of Communist troops who are killing my comrades and whom your husband is fighting are coming from. How we got into this mess is very simple. No guts. But it backfired, as the future history books will tell. *(This was written in January 1963!)* And all because we are fighting in Vietnam the same fight we bumbled around to try and avoid in

Laos. Here we are fighting the war in South Vietnam, fighting the war in the name of South Vietnam; but we are also fighting many of the same troops we would have had to handle in Laos, plus the Viet Cong of Vietnam. And we are fighting these characters under the craziest set of restrictions ever. *(Again, I mean President Kennedy and the small clique of advisors who micromanaged the war. We, the fighting commanders and advisors, knew more than they did, not only about what was really happening, but how to deal with it. But you only spoke up if you wanted to get in trouble.[4])*

We beat the pants off a VC unit and they run into Laos to escape and recover. Yesterday we discover a large force coming out of Laos, and as soon as they find out we're getting set for them, they head back for sanctuary until they figure the heat's off. This sort of thing can go on and on—and it has, since 1946, when the French had to start playing the game. In '56 the Viets took it over, and in '61 we decided to come in and try our hand. And here we are, the American GI, the final victims of our national brilliance. What makes it even more ironic, the enemy is firing American weapons at us too. These they got in a variety of ways, mostly by capturing them. *(In the mid-nineties, while doing research for this book, I also found out that 140,000 battle kits had been turned over to Ho Chi Minh by the United States government at the end of World War II, and these could have been part of what we were up against.[5] Also at some point around this time, I don't remember exactly when, and I wasn't able to find it in a letter I sent home, small teams of men equipped with radar devices looking like small portable searchlights came through Kontum to be deployed somewhere on lonely mountain outposts to watch the Ho Chi Minh Trail with their technology. I don't know who they worked for in our multiplicity of chains of command, or who benefited from their labors, or even if the blighters ever got back out safely.)*

Maybe the United Nations is the organization with the full responsibility to come in here and straighten out the rules of this war. It should be their job to see that it is being conducted above board. The ICC (International Control Commission), which they sent here to supervise

---

4. Andrew F. Krepinevich, Jr., *The Army and Vietnam* (1986), especially chapter 3, "Into the Quagmire," and p. 80, "MACV's Adviser Problems."

5. L. Fletcher Prouty, *JFK: The CIA, Vietnam and the Plot to Assassinate John F. Kennedy* (1992).

this thing, has been by and large a defunct joke. Even when all their latest findings condemned the Reds, no one listened or took action.

I don't write these words bitterly. I am too much involved in this thing to become bitter. *(The bitterness came later, when I returned home. I was abused, I had my military career cut short, and I ended up with no veterans' benefits.)* I need all my mental powers helping Captain Minh to keep this critical province secure. The task is to attempt to improve the troops while at the same time destroy as many enemy as possible, keeping our roads, towns, and key areas secure, and keep on pushing the enemy back into Laos-Cambodia.

The American officer is trained to destroy the enemy and to go after him until it is accomplished. This is certainly a new concept. Destroy him under a set of rules that restricts you but gives him a free hand. I can see that it is going to be frustrating indeed meeting the same enemy units over and over again, trading blow for blow and having to push them to their safety zone over and over again. The question is, who wears out first? I won't be around to find out, I guess; I'm a lucky one. *(Each of our twenty-six rifle companies had a TO&E [Table of Organization and Equipment] of 190 men and 5 officers. By the time I left, in 1964, the typical company had 130 men and 1 very green officer. Also, we were limited to Kontum Province, like the Missouri National Guard would be limited to the manpower pool of Missouri, but the enemy could go back to his sanctuary with impunity and bring in endless replacements from the north. When he was rebuilt and refit, he could come back at the time and place of his choosing. It should have been obvious to our leaders that we were violating the Principles of War at the outset by not protecting the flank in Laos. Or, why not just withdraw the Yanks?)*

We have other problems too. Such as when our Viets *do* get the opportunity to smash an enemy unit but then they let the quarry slip out of the bag. These guys [the Civil Guard] seem to be real erratic fighters. They can fight tenaciously as hell, but the leadership lacks an aggressive hunt-and-kill spirit. They are really short on officers and NCOs, and those that they have have no uniformity of ability. *(One of our huge assets in the American military system during that time was depth of uniform leadership quality. That did, however, make it harder for us to understand the Vietnamese problem, and it resulted in some bad analysis and poor decisions on the American side. It came down to what some might call a culture gap.)* My own "battalion," which is larger than some ARVN regiments (brigade size!), has less than one third the number of officers and NCOs they should have. Some companies are without officers and have only five NCOs! 104th Company lost their only officer in the Dak Sut fight last week, and one of their precious 6 NCOs. A company should have 5 officers and well over 20 NCOs. This is a big problem. Weapons and money we can get. Recruits and soldiers we can get—not easily, but we can get. Officers and noncoms and any kind of leaders, however, you just can't manufacture. It takes a certain personality type that this country does not abound with. It takes years of training and experience. And when you are short to begin with and you are at war, you lose the good ones faster than you can bring up new ones. *(Top American leadership at this time never grasped this. When they started playing games with the top leadership of South Vietnam, some of us knew at once that they were making yet another fatal error.)*

In time, this will be a first-rate army (ARVN), but we have a few more years of this situation to live with before we get there. We are just going to have to "gut it out," as we say. You adapt yourself to this situation or you go bats. Simple as that. The solution to the problem is so simple it's frustrating. The tragedy of it is that we have to build these leaders while they are fighting. This really makes it hard. They get into bad habits in combat and they survive, so they figure that's the way to do it. Boy, is that a hard attitude to change. At Ft. Benning when a guy goofs up in a mock battle they can stop the problem and set the guy right. You can't do that in combat. Worse yet, if the guy goofs up and they get through the fight alive (whatever the outcome), they think that's proof enough that they are tried and trusty troops. Teaching a bunch of people who speak another language under those conditions (actually three languages: Vietnamese, Bnai [Bahnar], and Jarai) is like ramming into an underpass on the Mark Twain Expressway.

Kontum, 3 February 1963

I was out on a two-day clearing operation near Kontum yesterday and today. Very interesting, but no VC. Last week a squad (up to 12 men) of VC crept up on the Tan Phu [Tân Phú] work camp, fired a few shots, threw some rocks, and otherwise created a disturbance, scaring people throughout the night. Also the same week, a small VC unit worked its way up to the edge of the outworks at Pleit Char outpost in the same area and started to dig in and create a disturbance. My counterpart, Captain

Minh, decided to try and catch these VC in a quick clearing operation. And so, out in the bush we went.

Well, at any rate, the operation, which I shall call "Operation Nam Dinh," started on the evening of the first. We used a battalion minus (an understrength battalion-sized force), which we divided into three separate groups or task forces, plus headquarters for control. They were organized as follows:

Task Force Nam Dinh 1 = 1 Civil Guard Ranger platoon
1 SDC platoon

Task Force Nam Dinh 2 = 1 Force Populaire platoon
1 Civil Guard rifle squad

Task Force Nam Dinh 3 = 1 Force Populaire company(-)
1 sector Ranger platoon

Task Force Command Group = Weapons Section, Headquarters
Company (Fu Le)
Ambulance Squad Command Group

The country is pretty hilly there and covered with thick jungle. There were swamp areas that were the only open spots. This being the dry season, movement over the streams and swamps was possible. But the big rivers to the north and west were too wide to cross. The area of operations was almost 10 miles square, way too big for close coverage. TF 1 [Task Force 1] was the blocking force in the west from line Plei Veh to Plei Hiep. Task Forces 2 and 3 worked westward from Route 14. We especially checked out certain areas that might be refuge for the enemy. On Friday night, February 1st, the troops in TF Nam Dinh 1 sneaked out of Pleit Char OP [outpost] and the western-most villages and occupied the high ridgeline in the west. They were to be the blocking force. They just dug in along the high ridgeline to the west and hoped the rest of us would chase the VC right into their guns.

Just as the sun peeked over the horizon on a nippy Saturday morning, the 2nd of February, a ground fog clung to the terrain all through the area. Just at this time the troops from Task Force Nam Dinh 2 clambered out of their trucks, which had been pretending to head south towards Pleiku. Faster than I can tell it, Nam Dinh 2 disappeared into the jungle and started pushing swiftly westward, while the trucks rolled on southward. At the same time, Task Force Nam Dinh 3 suddenly marched out of Tan Phu and whisked over the top of Route Nationale 14 in the ground fog, and they too were swallowed up in the jungle.

While these things were happening, Captain Minh and the Command Group and I were leaving the barracks area in Kontum to head for the first command post at the village of Plei Kolet. We timed our departure so as to allow enough time for the troops of Nam Dinh 2 to clear the area before we closed in. We had a considerable little fighting force ourselves just in case.

About 0900 we pulled into Plei Kolet amidst an ominous silence and emptiness. A mortar and machine guns covered while soldiers of the security section flitted from house to house and into the jungle beyond. I was getting a bang out of it, back in the saddle again. Once it became obvious we were not interested in shooting up the place, villagers began to appear from here and there until life looked back to normal again, except for the soldiers and weapons set up alertly about.

Captain Minh and I indicated we would like to parlay, and two head men came out. A soldier unrolled two bamboo mats across from each other, and Minh and I sat down crosslegged and faced two stony-faced chieftains who also sat cross-legged. We had a soldier who was an interpreter with us, and he sat cross-legged at the head of the group near me. (This was either a Bnai [Bahnar-speaking] or Jarai[-speaking] village.) First, Captain Minh told them why we were there, what we were doing, and that we would cause no harm and buy what we took. Then he proceeded to ask questions about Cong activity in the area, and the headmen seemed quite willing to supply information. VC had entered and left the village recently, from the west. One night, some time ago, they came in and captured a government worker who had brought food to the villagers. They took him into the woods nearby and killed him. No, they didn't know where the VC went; they were afraid to follow.

Begrudgingly, they agreed to sell us some rice and some chickens. We offered a good price, so I leaned over and asked Minh why they were so reluctant to sell. I could see chickens and all kinds of livestock all over the place. I had to push off an affectionate old sow in a rather undignified manner during the powwow. The story is this: The more livestock the village has, the wealthier it is considered to be. So many villages will actually face famine before they will kill and eat their livestock. But Plei

Kolet was not so primitive as this. Reluctant at first, when they actually saw us give the headmen the money, more and more food on the hoof began to appear, and soon our soldiers were seen chasing squawking chickens everywhere. I kept a sharp eye on the proceedings, first to make sure the natives got their money and second to cast a wary eye on what I was going to be eating. They (the soldiers) gutted the birds and plucked them (after a fashion) and cut off the legs, but unfortunately, everything else we ate.

We passed the morning taking radio reports from Nam Dinh 2 and 3 and passing our orders. Maps were getting marked up and copious notes were taken, but by lunchtime the boys out beating the bush reported no enemy in sight. I ate lunch like a true Montagnard-Viet, chopsticks and all. By now I don't take a backseat to anybody with the chopsticks. *(I could even eat soup with chopsticks!)* I looked longingly at the US Army C rations I carried with me, but I didn't dare offend my comrades, so I ate hearty with them. It wasn't bad. It wasn't the first time and far from the last. I'm not sick yet, so I guess I made it OK.

After dinner we relaxed for a bit in the village bachelor house [a rong house], where the natives, growing friendlier by the hour, invited us to set up our CP [command post]. You know what the bachelor house looks like from the little models I sent home. I sat on the thatched bamboo floor with my back against the wall and my automatic weapon across my legs and dreamed about my beautiful bride. What a peaceful little Southeast Asian village. If we could live in a dream world for a little while, I would have liked to see everyone PUFF! Go away. And suddenly my beautiful bride would appear, and . . . PFFFT! Back to reality.

At 1600 hours we quit Plei Kolet and move the CP into the walls of Pleit Char outpost. Our firepower really beefed up the already considerable armament of the post, so I felt even more secure than in Kontum. Pleit Char outpost is built in the shape of a triangle. The place is surrounded by a deep moat with cruel-looking spikes all over the bottom. The only way in is over two precarious boards thrown over the moat, which you have to balance your way over to get in. These can be pulled up in a hurry if attacked. The place isn't too comfortable, but I must admit it's a good battle fort. Captain Minh designed it himself. Each point of the triangle and the inside center has a powerful main combat bunker with weapons whose fire interlocks for maximum cover. Down each wall are machine gun bunkers.

Most of our outposts were triangular in shape, made out of wood and earthworks, with a strong bunker at each angle. The corner bunkers were equipped with machine guns providing interlocking fire with the other corner bunkers. The outpost was surrounded with mines, barbed wire, claymores, and panji stakes.[6] In the center was a command bunker equipped with radio and other automatic weapons that could sweep the interior of the fort. From here the commander could direct the fight and stay in touch with us at higher headquarters as we prepared a relieving counterattack. These outposts and the whole situation reminded me of my readings of how things were during the American colonial period, particularly during the French and Indian War.

We ate another chopsticks meal there and made use of the fort's communications facilities to maintain contact with our units. Still no contact with the enemy. (We had too much terrain to cover and too few troops. And then there is the matter of possible security leaks.) Nam Dinh 2 had reached Objective B by 1300, spent the afternoon scouring the surrounding area, and set up all-around defense on the spot for the night. Nam Dinh 3 did the same at initial Objective C. Nam Dinh 1 waited in vain at the blocking position. At Pleit Char outpost the night grew chilly. I climbed into my sleeping bag and fell asleep on a hard straw mat.

Morning brought more rice and chopsticks and still no VC, although all units were now thrashing about in the jungle. (Another question: Were they doing what they were reporting? I think so, but Nam Dinh 1 left its blocking position and pushed eastwards toward the Final Objective. Nam Dinh 2 and 3 pushed west to meet 1 at 0845 hours. The radio crackled, "Nam Dinh Hai, Nam Dinh Hai, day fou le, day fou le, o dou, lam zi cho ta lai?" ["[?] . . . ở đâu, làm gì cho ta lại?"] But no VC.

---

6. Claymores were derived from a German World War II predecessor called a shoemine [Schü-mine], which was a rectangular unit the size and shape of a shoebox that was set up on a wire frame. When set off, it worked on a "shaped charge" principle, blasting a wide swath of lethal pellets to the front. Panji stakes were small, knife-sized, sharpened bamboo stakes, hardened by fire and sometimes poisoned. These would be planted by the hundreds or thousands, shoved into the ground facing the enemy. They would sometimes be camouflaged or put in a pit trap or moat. One trick was to set an ambush on the opposite side of a trail where you had camouflaged some panji stakes. You open fire and the enemy reacts by diving right into the panji stakes.

At noon the thing ended, the weary troops started reappearing out of the jungle, and trucks rattled up to take them home. Everyone seemed in good spirits, even though the Cong gave them the slip. Couldn't blame them, I suppose. People would have been hurt in a fight.

But the reason the enemy got away bothered me for a while. I have been away from this stuff too long to have it at my fingertips anymore. So I had to more or less observe what went on to get my feet back on the ground. The answer now seems obvious. For the small force, we tried to cover too much territory. We left the Cong too many escape routes. Our only hope for a fight was bad luck for them and an accidental meeting. Minh's plan apparently was to scare them west with Nam Dinh 2 and 3 against Nam Dinh 1 in the blocking position and crush them from two sides. He left too many holes open. His forces were too weak. The area was too big to effectively cover. He should have picked an area he could seal off and effectively sweep in one day. As it was, the enemy could sneak out in any direction, and since we had to stop halfway finished at night-fall on Saturday the 2nd, the enemy had complete freedom to escape that night. This they no doubt did, and not a great deal of skill was required. For this type operation, we should have taken the smaller 1/3 of the area and sealed it off and swept it quickly in one day with the forces we had available. We might have had a fight or two then and bagged a few VC.

Next time I'll know better. However, my good friend Captain Minh is a 44-year-old veteran of many campaigns who fought long and well with the French. How to get the message across to him is really the biggest problem. He is not anxious to receive constructive criticism from a much younger American captain. How I get to him is a problem I have yet to solve. *(Typically Americans did not get to stay around long enough to really make an impact.)*

Here are some notes from Pleit Char: No medic, no bayonets for carbines, firing positions good, good firepower, no first aid pouches, too many personnel around cooking area, dirty dishes, no latrine in walls, direction of fire arrow not good, no evidence of onsite training, more patrol activity, water supply?

My attitude toward the enemy was one of absolute ruthlessness during fighting and then maximum compassion when it was over. My Vietnamese compatriots could not switch to compassion when the action was over. So I would go stand by the enemy prisoners for a while to protect them from any abuse, if I could. They had no clue that an American captain could speak Vietnamese in the northern dialect, so sometimes I overheard interesting things.

Fear came and went unexplainably. In some situations where I should have been thoroughly frightened I operated fearlessly, in some kind of euphoria, hoping to earn some high decoration or to do exploits. On other occasions, I would suddenly become scared and have to fall back on my paratrooper's discipline to appear to operate coolly.

Somewhere in this timeframe a couple of Australian officers visited us and followed me around for a few days to get the hang of things. There was a thirty-man jungle warfare specialist group in Vietnam in 1962 for training Vietnamese troops, and in April 1965, Australia committed an infantry battalion to the fight, I believe. Later in 1965 they reinforced it with artillery, air reconnaissance, and engineer troops. By 1969 the Aussies had almost 7,700 troops in Vietnam, and by 1971 they were all withdrawn. Like the US, they had conscripted troops; in contrast to the United States, Australia rotated entire units into and out of Vietnam, rather than individuals. America's rotation policy is something I will discuss later. The Aussies lost 386 killed and 2,193 wounded.[7]

8 February 1963

This situation, at times . . . it's worse than a jail sentence. You are not only a mental and physical prisoner, but you are driven relentlessly. The punishments are great for (perceived) failure. *(Not to mention that you are constantly at risk of death or maiming. We couldn't win the war that way, so the government should have modified its policy to win or brought us all home.)*

Realizing what we were up against, I had the temerity to suggest a secret operation against the enemy airbase at Tchepone [Xépôn] in Laos, where the Russians were reputed to be flying in supplies that were then being brought down the Ho Chi Minh Trail through Kontum Province. I wanted to take my maneuver element—about a battalion['s worth] of light infantry—and just sort of disappear into the jungle. Some days later, we sort of reappear in a surprise night attack against Tchepone and destroy the place. Then we would sort of disappear back into the jungle and reappear back in Kontum later.

---

7. *Webster's New World Dictionary of the Vietnam War*, 30.

I was chastised roundly for the idea at the time.[8] Only in recent years have I read that our government had different chains of command, like the CIA was supposedly involved in stuff like that. I thought army commanders were supposed to be running land wars! The army belongs to the People of the United States. Who were these other guys running parallel wars, and who were they ever accountable to?

In June of 1963, Secretary of State Dean Rusk and Secretary of Defense Robert McNamara proposed their three-phase action program in Southeast Asia "not as a contingency response to Communist tactics . . . but as a method of influencing the overall situation."[9] President Kennedy approved the program's modest first steps, which authorized US advisors to encourage right-wing and neutralist offensive military action against the Pathet Lao. Supported by CIA–Special Forces teams based in South Vietnam and Laos, the operations were aimed primarily at controlling approaches to Thailand and the mountain trails into South Vietnam. (Like I said!) The president also sanctioned air strikes by T-2Bs armed with napalm. According to a State Department directive to the embassy in Vientiane, the objective was to be able to meet new Pathet Lao attacks with prompt counterpunches in order to convey the message that PL may no longer instigate such actions with impunity. But during Kennedy's presidency, US advisory and covert activities in the Laotian panhandle never amounted to much.[10] What they did do was too little, too late.

Sometime in the mid-sixties in a conference on Guam with President Johnson, Vietnam's Prime Minister Ky [Nguyễn Cao Kỳ] suggested sending South Vietnamese troops into Ho Chi Minh territory to set up a base and put a thorn in the enemy's side. President Johnson said no.[11]

---

8. *Webster's New World Dictionary of the Vietnam War*, 389. Tchepone, in Laos, was a logistical center about forty miles from Vietnam on the Ho Chi Minh Trail. I was aware of its danger to us in the early sixties, but it wasn't until 1971, when the ARVN launched Operation Lam Son 719 against Tchepone and the Ho Chi Minh Trail, that I saw the military recognize its importance. This all would have been much easier to deal with during the 1961–63 timeframe. Some might say, "Oh, we couldn't do that." To them I say, "Fine, then be wise enough to see we couldn't win and get us out."

9. Memorandum from Michael V. Forrestal of the National Security Council Staff to President Kennedy, June 18, 1963.

10. William J. Rust, *Kennedy in Vietnam: American Vietnam Policy, 1960–1963* (1985), 89–90.

11. Nguyen Cao Ky, *Twenty Years and Twenty Days: How and Why the United States Lost Its First War with China and the Soviet Union* (1976), 100.

[Here is] an example of smart use of artillery. A certain base camp was supported by a firing section of 22nd DIVARTY (Division Artillery) 105s. These two 105mm howitzers could reach out to about 11,000 meters. Patrols from the base camp working beyond that range were being attacked by VC main force units, so one night our guys switched out the 105s for 155 howitzers, which could reach about 20,000 meters. Then the patrols went out beyond the 11,000-meter range of 105s and Charlie attacked. What a surprise when the 155s came crashing down on them!

The tragedy of war could really hit you sometimes. The Viets must have kept wooden coffins handy for the KIA [killed in action]. One fine sunny afternoon, Capt. Minh and I were returning to Kontum in a jeep after checking on something up north. We came upon one of our trucks stopped by a village. We saw that some of our men were up on a hillside at the edge of a village, helping a lone widow bury her husband, one of our recently killed comrades. The man must have been killed in an earlier fight, put in a coffin at one of our outposts, and delivered, on the way back, for the funeral. Captain Minh and I walked up to the graveside, snapped to attention, and saluted. Then we expressed our condolences to the poor woman and departed. I thought, This could have been a national guard outfit back home, outside of Janesville, Wisconsin, or Maryland Heights, Missouri. Thank God for our blessings.

Kontum, 20 March 1963

A gift from the Vietnamese and Montagnards I've worked with . . . the most beautiful bachelor house I ever saw owned by an American. It's as beautiful as any I've seen, in fact. The Vietnamese frequently give these little model bachelor houses to departing Americans. They always send here to Kontum for especially nice ones to give to generals, etc. The one they gave me was the nicest one I ever saw. Truly a beautiful work of art. It's even nicer than the ones I saw which Generals Taylor, Timmes, Harkins, and Eggleston received. That means something special to me. There also is a fine-looking model of an old-time Vietnamese weapons rack, which old-time Vietnamese warriors used to hang up their weapons in.

There was a nearby Special Forces camp where we had a team train-ing a bunch of Montagnards. These guys were hungry for US-style food, which I could get, and they had a lot of exotic weapons, among which was a Swedish K, a nifty little lightweight 9mm burp gun. We did business.

At that camp we were joined by a young doctor, a recent Harvard graduate who took advantage of every opportunity to come out to the Montagnard villages and minister to the sick. For this to happen, we had to deploy at least a platoon of troops around the village to protect it while the doctor and the medics worked. As I mentioned in an earlier passage about Doc Smith, soap and simple remedies did a lot of good. I used to get angry at the hypoc-risy of the pro-Communist left-wingers and the protesters at home since I lived through situations like that, where we had to defend the local people from the Cong so the doctors could help them. That Harvard doc was one fine, compassionate man. I believe it was a good thing for the Cong that they left us alone on those occasions. I used to have bad thoughts and fantasies about American pro-Commies and the Cong attacking us when we were ministering medicine in the villages, and how we, in righteous anger, would rise up and smite them, the enemies of the Lord. They made movies about stuff like that years later.

Kontum, 23 March 1963

Had a rough week. Had to tote a 70lb full field pack through the jungle mountains the last few days. The heat was fantastic. It was what I call a "flying patrol" (a long-range combat patrol). Only we were not "flying." We were really dogging it. We took about 20 men and tried to snap up some VC in the area between Polei Krong and Plei Kling. Then we reinforced the garrison at Plei Kling outpost while the boys out there thrashed around in the bush for a while in the valley up to Plei Kobay.

The company commander at Plei Kling was the best officer we had in the 10th CG next to Captain Minh. After them, and possibly Lt. Thich, the exec, there were few officers and no good ones. I overheard Thich in the next room say to the company commander, "You remember Cap-tain Frankwicz, don't you?" The swashbuckling CO replied, "Hell no. All these white guys look alike." I tried hard to stifle a roar while Thich replied, "Shhhhh. Careful! This guy speaks Vietnamese."

Shortly after, we heard a shot. One of the troops on guard on the wall shot a deer. They cut the flesh into small slivers and laid it out on a mat, raised up by stakes to dry in the sun. Had some from a previous kill for the evening meal. It was great.

Yesterday we marched back in through terrible heat. The heat waves just bounced off the ground. It looked just like the heat waves you see coming out of a toaster. At one point we took a short break at a place where the mountain just dropped off at the edge of the trail. A sniper on the next mountain over started popping away at us, but he shot high, and the rounds snapped through the leaves overhead. *(The enemy can identify our command group by the radio antennas near us, so we are the prime targets.[12])* We were too exhausted to move. He either got me or he didn't, sorry. *(That's one way people become casualties in war: too exhausted to take cover. Fortunately, I got through that one so six more kids and a bunch of grandkids could be born.)*

When we finally reached the ferry site at Polei Krong (over 3 mountains), we just fell into the river, equipment and all. Man, did that cold water feel good! We were unbelievably filthy. I had actually worn a new pair of jungle boots on my feet on the way out. They were broken in just before we left. [By the time] we closed in on Plei Kling, the soles were split and I was in danger of beating my feet into hamburger. Our second element was radioed to bring me up some new boots, and I had to break them in on the rest of the march. Talk about sore feet! My tongue was hanging out so far it was getting sunburned, and I had to keep from tripping on it. And such swarms of bugs you never did see.

The food out there was horrible too, except for the deer jerky. They had one dish that was really something. Congealed blood pudding! Yes!! They came up with bottles of white lightning or rice moonshine and they said, "Drink this, eat this, very good!" I had already eaten just about everything they had to offer, but this was the end. I declined as politely as I could. When I get home, I suppose that I will let myself think about all this stuff and really lose my appetite. Right now I don't think too much about the nasty things. We still manage to keep smiling and laugh this stuff off, so we're still doing OK, I guess.

---

12. One or more radio operators who were back-packing field radios with long whip antennas would follow close to the commander in a command group. The enemy could locate the commander from the whip antennas, which is one reason why combat arms, particularly infantry officers, suffer heavy casualties in war.

(Most of my fellow Yanks caught dysentery while in Vietnam. I escaped that but had dysentery for nine months after returning to the States. Finally, I ended up in the hospital with some kind of stress-related stroke. I thought it was probably combat fatigue that had caught up to me because of the nonstop high-stress pressure.)

A few days later we brought a higher headquarters group in a Huey chopper formation out to visit my swashbuckling buddy. Suddenly, after being out for a while, the command pilot announced he was lost. I climbed up behind him and was able to find a reference point and navigate us in. The maps were only good near the roads, and beyond was cartographic guesswork. I still have my battle maps. I felt proud that I had really earned my Expert Infantry Badge and could pick up in the middle of nowhere and guide us in, either from the air or under the jungle canopy. *(This was in the days before GPS.)*

Kontum, 30 March 1963

The night Plei Mrong got hit I was safe in Kontum. When something happens in the war, it happens at one little spot on the map for a matter of a few hours at the most. Even if many things happen at once, [for] the vast majority of territory and the vast majority of time, over 99% [of the rest is] safe and free of activity. Some people back home may think of war more total than it is. And a lot of guys like people to think that way because it makes them look like rough, tough guys that have really gone through hell. This is no picnic, don't get me wrong. But neither I nor anyone else *(at least in that time and place)* is in constant danger or in total calamity of war, and when we are it's not half so bad as the so-called heroes, the newspapers, and Hollywood want to make it.

So far this one can't touch the nastiness of World War II and Korea, as those wars were for infantrymen, marines, and many others. We go into the same situations and we need the same courage, but so far, in the end, while it can be miserable, it's not as rough, not as dangerous, not even as bloody. I think I might be awarded the Combat Infantry Badge, but we didn't go through any total hell for extended periods to get it, like some of our forebears. I've read some of the stuff they've written about operations I've been in, and they make it look like we really went through it. *(I had done fifty-some combat operations over the year, from resupply operations to clearing operations to long-range combat patrols, and there were no firefights worth the mention. Later, when American troops were more fully committed, it became as nasty as any war we've been in. It was our hope that our service would prevent the need for others to come later. But the government did not effectively use us.)*

Plei Mrong was over the mountain from us, south on Route 14. There was a Special Forces training camp there. Enemy agents had infiltrated the camp and cooperated with a VC night attack against the place. While the troops were in their firing positions shooting at the charging enemy, the inside agents ran around behind the firing lines shooting our men in the back. The result was a quick disaster. The VC got in through the gaps, and a massacre ensued. The small group of American Special Forces advisors and a few others got into the center of the camp and sat down in a circle back to back, firing at anything that got close to them. Over the mountain we could hear it and see the lights like distant lightning from the intense firefight.

Relief came as soon as possible, but the place was virtually wiped out. It was just out of our sector, so we came down the next day as observers when the cleanup was going on and we saw what happened. These guys got a firsthand idea of what Custer must have felt like, or maybe the boys in the Alamo. I believe the fight resulted in a front-page story in one of the big magazines back home, possibly *Life*. The guys in that one knew they were in a fight.

There was a problem with Special Forces deployment. They were supposed to be a mobile force training guerillas. But as my friend from Germany, General Harold K. Johnson, wrote, "What they did was build fortifications out of the Middle Ages and bury themselves, surrounding themselves with concrete; and simply building little enclaves in each tribal area would not have very much utility as far as creating an environment for free society was concerned." Basically that's what they had in my area, little forts like our Civil Guard and SDC had. The Vietnamese Special Forces were ill-equipped to assume the responsibilities of their American counterparts. They were poorly trained, incompetently led, and insensitive to the needs of the population. Col. Wilbur Wilson said, "These forces essentially man bases which are in effect static defensive positions. Operations conducted from these are no different from the operations conducted by a typical infantry unit under the same circumstances." [13]

---

13. Krepinevich, *The Army and Vietnam*, 74–75.

Kontum, 6 April 1963

We were out to Khon Brai and Plateau Gi the last few days. The last time
out there. One more trip into the bush on Monday and that will be all.

Kontum, 10 April 1963

I came safely out of the field for what should be the last time this morning. The mission was sort of a crowning accomplishment for us, and I
feel rather proud of it. We strengthened our positions in the north and
moved our companies (or four of them) into vigorous patrol activity
blocking off the dangerous Bong Hong Zone of the Viet Cong. The
outposts are outposts no more but rather bases for patrol operations. We
have come a long way. Now ARVN *and* the Civil Guard are carrying the
war to the Communists as never before. We've had some pretty rough
stuff going on up there, and it's getting rougher. It's a good time for me
to get out. I'm pretty tired to go on anyhow and starting to get the jitters.
Good luck to Van and the newcomers. It's their turn now. We've done a
job that I think we can be proud of. Things were a lot different when we
first arrived [compared to] the solid progress that's marching along now.
But the Cong isn't exactly sitting down either. The slugfest will last for
quite a while.

We lost a B-26 bomber this week over the same area I was patrolling
a few weeks ago near the Yalu [River].[14] It was strafing and bombing
and had climbed to 3,000 feet to dive back in for another run when a
wing tore off from metal fatigue.[15] The poor guys, two Americans and

---

14. For specs on the B-26 bomber, see Karen Leverington, ed., *The Vital Guide to Fighting
    Aircraft of World War II* (1995), 29. This plane was designed in 1940 to replace the
    A-20 Havoc and was called the A-26 Invader, built by Douglas. After the Marauder
    disappeared from the scene, it was redesignated the B-26. This aircraft was used in
    Cuba and Vietnam by the CIA (and the Air Force?) from 1961 to 1963, and, like much
    of the stuff we had to use in the army, was obsolete inventory and a death trap. The
    government had a bag full of obsolete tricks to kill off their own boys.

15. Katz Keating, "Plausible Denial," 48–49. In writing about Operation Farmgate, Katz
    Keating wrote, "The crews in some B-26s and T-28s were dying as a result of what some
    euphemistically termed 'equipment failure.' In fact, the airplanes were falling apart in
    mid-air. *[cont. on next page]*

    "'These airplanes had been used in World War II and Korea, and they were tired,'
    Kittinger [a pilot] says. 'And we were using them as fighter-bombers.' The old airframes
    simply were not up to the task: 'The wings started coming off them.'

    "'If a wing comes off, you get just a violent roll,' Kittinger says. 'The G-force would
    preclude you from doing anything. You can't get out. You don't have a chance.' *[cont.]*

one Viet, never had a chance. She spun in and blew up in less than four seconds. There wasn't much left. It took our people a half a day to get in there. These old aircraft our guys are forced to use over here are a menace. They are falling out of the sky right and left. I can count over 20 planes shot down or crashed just around here. We have gotten so we'd rather walk through an area likely to be ripe with Cong ambushes than fly one of those old crates in. The H-21 helicopters you see all the pictures of are the worst. I don't have to fly in them anymore, so I can talk about them now.

I had forgotten until Hank Parkhurst reminded me one day that we took off on a mission with those things some time ago, and as we were climbing, the one we were in quit running and just fell back down to the ground. It was one hell of a jolt. Do you know what happened? All the other troops disappeared off into the distance in a large formation in the sky, leaving the leaders behind. They brought another old rickety one right in and off we went again. Needless to say we made it. I must have been terrified. *(The military wrote that we were "the best and the brightest," we who were selected to be the first big wave of advisors, but they gave us the oldest, most dangerous aircraft and the most obsolete weapons, and not a full complement of them. This cost us lives, maiming, and serious injuries. I don't understand the logic; I think it was Mr. McNamara's efficiency experts' doing. They used up the old stuff, all right, and got rid of a bunch of "the best and the brightest" too, whose names are on The Wall. My group is the first section right of center.)* Bullets go right through those things like paper. You never know when and where they'll start ripping through. They are all patched up like patchwork quilts where the bullets went through. I much prefer to slug it out on the ground, and luckily I've not had much of that either. God has indeed been taking good care of me. Somehow or other when I walked, drove, or flew into nasty situations, I

"In February 1964, after a number of B-26 losses, a wing failed on a B-26 during a demonstration at Hurlbut Field, killing two crewmen. The entire B-26 fleet was grounded.

"There was a brief journalistic outcry surrounding the problems with what Farmgate crews irreverently termed the 'folding-wing version' of the B-26. Soon after the Hurlbut Field incident, *U.S. News & World Report* published some of [pilot Edwin] Jerry Shank's letters home, in which he complained about conditions in Vietnam. Among them was an indictment of the B-26: 'That airplane was a killer.' The letters were all the more arresting because they had been supplied to the magazine by his wife shortly after Shank had been killed when one wing of his T-28 sheared off during a bomb run."

was somehow protected from them, although I was ready and expectant of them. I have been very much afraid many times, I think. But I have always been able to do what had to be done and never turned away from anything.

13 April 1963

I was gratified and humble to hear that I received a very high efficiency report when I left. Since this was a combat efficiency report, its meaning was twice as important to me. I'm pointing towards [earning the rank of] major rather nicely right now while still a young captain. This news that I seem to have come off so well over here certainly affects my thoughts about remaining in the army. Rumor had it that I was probably destined to be a young major and, furthermore, my name had been submitted to attend Ft. Leavenworth Command and General Staff College not long after Benning (where I was in the Advanced Officers Career Course). That truly confuses me. They send people they think might make general someday to that school. And very few OCS [Officer Candidate School] graduates. Being a young captain has been very hard on me. At the moment I'm more interested in being a family man.

Kontum, 28 April 1963

The first two weeks of last month the VC came down out of Laos and the mountains and started attacking villages. They took great sport out of running over men, women, and children of any age. The purpose was to break up the strategic hamlet program by destroying the faith of the people in the government's ability to protect them. The VC started well. They overran two villages before our people knew what was happening. When our troops did get out, there was a river between them and the enemy, and we lost some good men trying to get at them. When the enemy did pull off, we managed to bring artillery up and pound the hell out of them. A week later they came down again and hit, but this time it was a give-and-take fight and the villagers (Self-Defense Corps) held out. The third week they came *again*. A confused fight developed. One nine-man squad from 897 Company was able to get to a village and attack the Cong. The Cong had over 100 men but for some reason tried to avoid our troops. (Incidentally, our units were made up mostly of Montagnards from this province, who were defending their own homeland.) The Cong

mixed into a big crowd of villagers, so our men had to charge in with bayonets to ferret them out and kick them out.

That was it. We pulled together every trooper who could walk from
four companies, 897, 957, 104, and 105 Combat Companies. ARVN
added Rangers, some 40th Infantry troops, and the blessed artillery. Into
the hills after them we went. Our troops did OK.

*As of 30 June 1963, 10,200 army personnel were reported in Vietnam,
including 15,400 US military personnel in total. Only 1,200 to 1,600
of us were combat effective—that is engaged with combat units. [Editor's
note: According to updated records, more 2.7 million Americans served
in the Vietnam War. Of those, more than 58,220 US military personnel
were killed. From 1964 to 1975, the total number of war-related deaths
in the region—military and civilian, on all sides—likely totaled more
than 3.5 million.]*

CHAPTER SIX

THE END OF MY ARMY CAREER

JUNE 1963 AND BEYOND

AS THE READER can see from my letters home, by mid-1962 I had started to realize that my military service wasn't going as I'd hoped. My general thought was, "These guys (in the Kennedy administration) are going to kill us!" As a professional soldier, that's the way it goes, you might say. HOWEVER! As an American professional soldier, I trusted the government to use us wisely, for a righteous cause, and with a reasonable objective. It became increasingly obvious as the months went by, though, that this wasn't happening, and even after returning home it looked less and less likely.

I'd also been recently married when I left for Vietnam, and it began to dawn on me, "They are going to leave my wife and babies without bread!" There were probably plenty of soldiers of all ranks who should have thought things like that through too, and now their names are on The Wall, and it hasn't been easy for their survivors.

Once I realized where things stood and where they were going, I started asking myself, "Can I make a difference? *How* can I make a difference?" I could not have prepared myself better for going to Vietnam, I'd done my best while I was there, and I reported any grievances within the system, carefully "appealing policy" according to the discretion of good army politics, having learned from experience that it's best to appeal within channels, preferably verbally. I did what I thought was right, and it still didn't work.

The situation was compounded by the fact that I had some physical ailments when I came home. For the first nine months of being back, I suffered from dysentery. Most guys got it while in Vietnam, but while I didn't get it much there, I really got it back in the States. The military put me into a nonstop high-stress routine when I came home—first the Advanced Officers Career Course at Ft. Benning, where I managed to graduate in the top 10 percent, and then I had a job as Regimental S-4 (Supply, Logistics, and Finance) of the 2nd Training Regiment at Ft. Leonard Wood, at a time when they were rebuilding the post and everything was a big hassle. The American system of management could subject you to unbelievable periods of stress, and they didn't care or understand that there was a limit to what a man could take without a break. It was the thing to do, to really lay it on you.

But the biggest factor in my leaving the army was that, as my family and friends can testify, I foresaw what lay ahead for me: three to four tours of duty, fifty thousand deaths, a probably unhappy ending to the war, and with the possibility of my own death occurring anywhere along the line. Nothing would change. If I couldn't make a difference in the system as a young captain, chances are I would be labeled too much of a dissident to become a commissioned officer, when I might have more influence. So I decided to look after my family first (since I could no longer trust anyone else to), and I would try to appeal policy from the outside, where I had more options.

I resigned my commission in early June 1964, and getting into civilian life turned out to be difficult. I thought there were lots of management opportunities, but there wasn't much available for an ex-army captain with three years of college. I already had good management experience, but it didn't translate to corporate recruiters, so instead I took a training post with the St. Louis County Civil Service Commission for $7,500 a year and struggled my way through night school at Washington University, preparing for the computer field. I tried to join the American Legion to use that as a platform, but they were not recognizing Vietnam vets yet.

Meanwhile, I continued to share my thoughts about the US military with people I thought might be able to make a change. At one point I wrote a test case letter to my congressman, who forwarded it to the Pentagon. An Air Force colonel in the Pentagon read the letter but only refuted my case using the same facts and figures I had used to build *my* case. The congressman didn't help at all, merely forwarded the colonel's reply without reading it.

During this time, I was consumed with struggling to support my family and to re-educate myself for a new profession, so for the most part I had to set aside my crusade. The only time I shared my opinions on the war was when someone found out about my background. Some folks wouldn't believe what I had to say, in effect calling me a liar, and that led to some problems. Various types of abuse toward Vietnam vets also began to crop up here and there, lasting all the way into the 1980s.

But the real killer was the public apathy. I had lived through World War II and Korea, and I remember how we all treated the GIs like heroes, during and after World War II especially. I had put these men on a high pedestal and seen myself as a young lad with his own chance to join the services and move from poverty to what I thought would be a level playing field in the military. I had dedicated the first ten years of my adult life to being a warrior, and so it was a severe blow to everything I believed in and to my personal self-worth to be abused by segments of society and, even worse, ignored by the nation. We were mistreated by some, while most of the rest looked on apathetically. We were supported by only a few. The media, particularly on TV, got worse and worse. The dissident groups increased. Watching all this was almost more than I could bear. But I could bear it—I was a tough paratrooper—and I slowly built a fine career in computers and raised a large family.

I've led a good life, but I still resent that I just missed out on the veterans' benefits I had not only counted on but earned. I'd resigned my commission in early June 1964, and the government later awarded veterans' benefits to those who were in service after June 30, 1964. It's pretty evident that, when handing out benefits, they missed us early guys. Was it deliberate? I don't know. But it definitely wasn't fun thinking about all the rear-echelon people in, say, Honolulu who qualified for veterans' benefits, while those of us who had achieved more and seen more action were rejected because we'd gotten out before June 30. I wondered if it wasn't done this way to punish "the best and the brightest"—those of us who had sacrificed their careers to appeal policy because we thought the system was doing us wrong.

I don't know what excuses they could have for not giving us the benefits we'd earned. Modern computer technology was good enough by then to find the early veterans and reward them. And paying veterans' benefits to those of us who had seen real combat would probably not be resented by anybody, *and* it would be a lot less expensive than the other veterans' programs we have had in place. Maybe the best excuse I've heard is that there was a fire at the army's record center in St. Louis around 1975, and the condition of the records may not have been so good, causing some of us to be overlooked.

But what a blow. I'd gone in with the attitude that if I took care of business 110 percent, the system would take care of me. But in Vietnam we learned there were a lot of separate agendas, and even with the best of intentions, you could run afoul of someone real quick and get on the system's bad side. It didn't matter how good of a soldier you were. It didn't matter that you'd been in combat. Not just after but during Vietnam, I saw the entire worldview on which I had based my life, and for which I was risking my life and my family's well-being, revealed as terribly flawed. My fundamental belief system started to come unraveled. And I was not the type that could say, "OK, this is just a job. I've got X number of years to go to retire; don't blow it." Nor could I say, "I'm going to string along with this until I get my chance to change things." What a testament to this system to realize there were guys like us stuck in that situation while some rich-and-famous types in the US administration and military had done terrible things and still been allowed to walk free all those years.

Perhaps one of the main problems that has kept me thinking about this so many years later was that I was more invested than some other guys. I had started getting my mind into Vietnam, studying the situation there, in the late fifties, years before I joined the army. Then I went there and saw it all for myself. And then, on coming home, knowing what I knew, I still had to endure watching the war from afar until it ended in 1975, and *then* I had to go through all the postwar issues pertaining to veterans, some of which were pretty tragic. It didn't stop for me until the Gulf War. That was about thirty-two years, more than half of my life. Of course, others saw worse combat than I did, but they went in in, say, 1967, got out in '68, and only had to live with it from then. By comparison, by the time I set foot in Vietnam, I had already been involved for ten years and was enduring the agony of knowing things shouldn't be happening the way they were.

Each of us went through our own special trauma unique in the history of American war, and sometimes we were confronted by a hostile public, but always we faced an apathetic public. There were a few people who passed on expressions of patriotic and personal appreciation and support through the years, especially in the churches, and I appreciated that. And sometime in the mid-sixties I even saw a front-page newspaper article decrying the military's loss of twenty-eight thousand of my fellow officers who resigned with between three and thirteen years of experience. At the time I saw the article, only a few hundred could have been Vietnam veterans, but it must have been a deluge in the years that followed.

Did the United States suffer from those resignations? If most of us had stayed, it probably would have added a bunch of our names to The Wall, but it also might have subtracted off a few thousand others, *if and only if* the people in charge had used our experience wisely in combat. Instead, I had seven children and, so far, five grandchildren. [Editor's note: As of 2025, there were six grandchildren and seven great-grandchildren.] So many more soldiers like me disappeared into society, some with wonderful lives, I hope, and some with real problems that are still affecting them into the nineties. We all deserved better.

PART III

OTHER THOUGHTS ABOUT THE WAR

Both during my time in the war and after I came home, I thought a lot about why the war went wrong and what I might have done to prevent it. Chapter 7 includes three pieces I wrote during the 1960s, and chapter 8 includes three pieces I wrote for this book in the late 1990s.

CHAPTER SEVEN

WARTIME WRITINGS

AN ANALYSIS OF THE SITUATION IN SOUTHEAST ASIA

OCTOBER 1961

At the time I wrote this piece, neither the Kennedy administration and its military nor the American public seemed aware of or concerned with the threat of Communism. Richard Gid Powers, in Not Without Honor: The History of American Anticommunism *(1995), wrote of this period, "The extremism hysteria of the early sixties made public discussion of communism seem too inflammatory to be tolerated, and analysis of moral issues seem a sign of weakness in an administration where only toughness and rationality were admired. . . . One of the greatest ironies of the history of American anticommunism is that the decisions to involve the United States in Vietnam were made with so little reference to principles, values, or goals of anticommunism. It was ironic because, objectively speaking, the Vietnam involvement was nothing if not anticommunist, and because the ensuing catastrophe would (quite logically) be blamed on anticommunism, turning anticommunism into a scapegoat for national disaster."*[1]

Current events in Vietnam indicate several things to me regarding the situation.

1. We are playing the game exactly as the Communists predicted we would in their prior planning. They sized up the JFK administration, and subsequent events show the Reds correctly diagnosed the lack of willpower of the administration.
2. The American people have not been kept properly informed regarding the Southeast Asia situation. This is in regards to the information required by high-level officials for action and the press releases to the public. There may be some mitigating circumstances, but this basic statement nevertheless holds true.
3. From all indications based upon this soldier's analysis from news releases that are available, leaders and soldiers at the tactical level are performing very ably in the field.
4. There is no question in this soldier's mind that the situation in Southeast Asia is the result of apparent incompetence on the strategic level in our national leadership.

I fully realize that the above statements threaten the security and power structure of the powers that be and will therefore lead to accusations branding me as "radical right," etc. I also realize the urge for political self-preservation has caused certain quarters to "muzzle" the military, keeping it from informing the people of the results of military insight into areas of conflict with Communism.

Nevertheless, the war of the Communists against us is a total war embracing all the fields of endeavor of life. It has one common factor found as a cross-section throughout each of the areas of its totality, and that is the phenomena of conflict. In our Western culture, the one profession that deals with the most forms of conflict of this scale and type is the military. As a result of the required training, the military as a group has been in a better position than most other groups to regard the

---

1. Richard Gid Powers, *Not Without Honor: The History of American Anticommunism* (1995).

activities of the Communists with keenest perception. The realization of this fact to some degree in high circles in 1958 led to the executive order which allowed the military to use its generally superior knowledge as a group to fill the information gap which exists not only in the general public but, evidently, in many policy-making circles of the nation. The result of these programs caused certain actions and reactions that ultimately led to the "muzzling" of us in the military. As the public became more and more informed about the history and the inner workings of the Communist menace, the public began to ask the question "Why?"

This inevitably caused certain high circles to squirm and react with the instinct of political self-preservation. Their reaction was to attack any group which attempted to fill the "information gap." Coincidentally, the Communists are also attacking these groups for their own reasons of self-preservation.

Another result of these programs was to produce people who felt compelled to react to relatively radical or desperate measures to combat not only the Reds but also those who they thought to be incompetent policymakers in our own system. In any case, and however ideological or selfish the motives of these groups, the end result has been harmful to the nation. In my opinion, the information gap remains, and one of the best-qualified groups to cope with the situation, the military, are now silenced. The spirit of military-sponsored seminars was not one of providing political influence in the same sense as alleged by enemies of those activities.

The spirit of these seminars is to provide the military and the public with historical and analytical information concerning the Communist Conspiracy. It just so happens that the historical facts also provide damaging evidence concerning the competence of various power groups in this country.

The military were encouraged to assume the role of "teachers" as a sort of crash measure to fill the informational and educational void. It was necessary to improvise for the time being and use *whoever* was qualified for this instruction. This provides part of the background from which I shall launch my arguments supporting my four main points. My personal background, which stimulates me to make these observations, is one of over seven years of service as a professional soldier. I have made it my business during these past seven years to establish certain fields of inquiry for myself in order to increase my overall competence.

I have studied in some detail the field of guerilla and anti-guerilla warfare. I have conducted some troop training and classroom instruction in that area. To support these studies and to increase my competence as an officer, I also studied Communism to a fairly great extent. I became well versed enough to speak before several public groups on the subject here in St. Louis, before the order appeared which prohibited us from any further participation. This led me to still another field of inquiry. I studied the various factors in this and other countries that seemed to contribute, wittingly or unwittingly, to the successes of the Communists. I became well enough informed as an army officer to be able to analyze much of what is happening.

Confining my remarks to the Vietnamese situation, I was able to predict months ago exactly how the Reds would develop that struggle in relation to the overall world situation. Events there have progressed to the point where I was absolutely correct. I even worked up a private project for my own benefit attempting to devise tactics and strategy for an all-out anti-guerilla campaign against the Viet Cong. I abandoned the project as being too costly and generally pointless. But now I wonder . . .

Creative thinking is not encouraged among rank-and-file captains in areas outside of their sphere of responsibility. You "question too many people's competence," when in fact that was not your prime motive at all. The higher-ups also don't like that this kind of thinking creates extra work. These are two major reasons, I might add, why major creative contributions to the art of war are not forthcoming from American military men, or at least not to anywhere near the existing potential. Few people outside of jobs requiring creativity ever exercise it.

That is the background as to why I feel I have qualifications to say what I'm saying. Now I shall launch into a discourse on my four main points.

It should be fairly obvious to anyone who is well informed on Communism that my first point of the four was obviously true many times in the past. It was true again for the situation in South Vietnam. On the basis of the Communists' past experiences, and on their top-level evaluation of the new Kennedy administration, the Communists found they could predict our reaction to their moves in Vietnam and act accordingly. Actually, they sized up the new administration quicker and better than we the American people did.

Their lack of respect was especially evident when they dusted off the old plan they used against the French and used it again against us.

According to Joseph Alsop, Viet Minh regulars are in the fight, and one specific regiment has been identified as the 803rd Regiment of the 324th Division. If this is true, the Reds are even using the same cast that performed against the French—a real insight into the opinion they have of the JFK administration. Among their many brilliant campaigns against the French, this crack unit dealt one of the last decisive blows by virtually annihilating France's Group Mobile 100 in March 1954. This French unit was previously the French contribution to UN forces in Korea and was highly regarded as a fighting unit. The French put up such a gallant fight in a hopeless situation that they have been remembered by the folklore of that region in Vietnam, where the Communists' Regiment 803 destroyed them. GM 100 was written off the books by the French army in September 1954.[2] The lessons they and their thousands of fallen comrades left behind have not been wholly heeded by our top-level people. I would guess that this is because very few top-level persons in this country have studied the French campaign in Indochina.

It seems obvious that a definite discernable pattern has developed here. Guerilla operations by the Communists were kept rather quiet in South Vietnam during the Eisenhower administration. This may have been due to Red concentration on the Laotian area. But more likely, in my opinion, it was due to Communist fear of reaction by the Eisenhower regime. American reaction at the time in Laos was ineffective, to be sure, but simultaneous conflict in South Vietnam would have stirred up a big Yank reaction. This probably was not considered safe during General Ike's administration. But Red buildups were in full swing for the appointed hour. The Communist attack was in the first two of five stages of guerilla warfare. [See the Appendix.] The first stage is the creation of favorable conditions, and the second stage is the so-called incubation period. This, I think, was evident to those few who understood this type of warfare at the time.

The Communist signal to go all out for the seizure of South Vietnam occurred as soon as the Reds had a chance to size up the new Kennedy administration. There is no need to dwell upon the events which brought top-level Communists to those conclusions. Suffice it to say, the Reds thought, and so far rightly, that the Kennedy administration and the American people would allow Southeast Asia to go down the drain—not

---

2. Bernard B. Fall, *Street Without Joy* (1961).

all at once, but at a much faster rate than was possible under the Eisenhower administration.

Consequently, in spring of 1961 the Reds pushed their guerilla campaign in South Vietnam. Stages 1, 2, and 3 of guerilla warfare became obvious in the few news reports that filtered into the more remote pages of our newspapers. These stages appeared in various degrees and at different areas within the country, as indeed they still do. The creation of conditions favorable to revolution, Stage 1, was carefully tended to. In those areas where this was working, Stage 2, the incubation period, was ushered in. Opposition to the Diem regime and its connections to the US sprang up more and more. Local complaints were even more played upon. Sabotage cropped up. Beatings and assassinations on a terrible scale took place. Stage 3 was also carefully nurtured into favorable regions. Guerilla buildup, preparation, and minor raids occurred during this period. Then, as now, various regions and provinces were in different stages of revolution, as I have described. Apparently the Reds had imported their revolution, or the greater part of it, along with the necessary catalytic agents.

Because they felt relatively secure in their undertakings during the summer of 1961 (you may recall the void of information in the newspapers on Vietnam prior to and especially during the Berlin Crisis), the Reds operated a controlled revolution. They supply the impetus to speed it along wherever they can.

The Western situation deteriorated so rapidly in South Vietnam that by late summer 1961 the Reds had reached Stage 4 of guerilla warfare in many regions. This is the all-out struggle by the Red guerillas to swing the balance of power to the Communists. Only then, and possibly too late, did the Kennedy administration see the gravity of the situation in Vietnam. A few weeks ago the Reds apparently flaunted their initial move into the final phase, Stage 5, of the conflict. This at least is their announcement—that they consider that they now have the balance of power necessary to secure final victory. They have introduced regular forces into the campaign. This is the fifth stage.

Guerilla warfare cannot of itself attain victory. Guerilla forces must sap enemy power to the extent that the balance of power will ultimately permit full-scale, conventional actions. Guerilla forces slowly evolve into regular forces in any area that becomes desirable as the situation progresses and permits. In some provinces, notably the north, this was evidently accomplished by bringing in North Vietnam Regiment 803 and other as-yet-unidentified units. In other provinces, news dispatches report that they are definitely attempting to organize regular units from their guerilla forces. Some areas are probably not this far advanced, but the pattern is evident to any military student of this situation and of guerilla warfare.

According to Joseph Alsop, the Communists succeeded in maneuvering into this final stage in spite of a revolution by North Vietnamese in their own backyard. They are apparently successful because they take firm, deliberate steps to smash such movements. It seems the situation has very nearly progressed to that of January 1951, when Communist Commander in Chief of the Viet Minh, Vo Nguyen Giap, progressed into the fifth and final stage of the war against the French by leaving his Red Chinese sanctuary and waging his final campaign, which finally won North Vietnam for the Reds in 1954.

Did we of the West learn? No, we did not. Most Americans still think the French lost a "guerilla war." I dare say they will think Vietnam was lost in '61 or '62 to "guerillas." But the decisive Communist power from January 1951 to the French collapse—and from October 1961 until another Western collapse(?)—was not guerillas but regular forces. The difference was the Communist regulars were adapted to the terrain they fought in. The Westerners were not. To be sure, the Reds won their guerilla campaigns, as they have this one in 1961; otherwise they would not have evolved the fighting into conventional warfare in many areas, as they have. Up to the moment (October 1961), the Reds appear to have utilized the same pattern and plan they used against the French, but now against us. Only the circumstances and the situations have changed. Evidently, the Communist actors have not changed. Our counteractions, while more suited to the purpose than those of the French, appear to be failing. Too little, too late.

Thus the strategists have left the situation in such a state that it can only now be saved by a large-scale limited war or an outstanding military leader of a caliber we do not seem to have at present. Of course, the policy-makers of the land could write the area off rather than do anything. This seems to have been the current effect based upon a wait-and-see policy.

In the meantime, our military people in South Vietnam seem to have been doing a fairly good job at the tactical level. Small, unit-level, range-type infantry training has produced better results than the French

European tactics approach. But the high-level people have kept this activity at too small a scope to prevent the Reds from grabbing the balance of power.

The situation in Southeast Asia is very grave indeed. It seems logical that the US leadership apparently did not inform itself well enough to understand what was going on. When the handwriting did finally appear on the wall, the result has been a curious lack of reaction. If the American leadership was aware all along of the progress of events in Southeast Asia, due to this lack of action, then we can assume that despite propaganda to the contrary, we have tacitly written off Southeast Asia months ago and provided the Reds with a decisive victory, with very little fighting in every way. Whatever the case, it seems safe to make the basic assumption that if the government would not take steps to crush the Red drive in its budding stages, it undoubtedly will not do so now that the Communists are rolling downhill in a full-scale offensive.

BASIC PROPOSALS FOR BETTER UTILIZATION OF THE CIVIL GUARD AND THE SELF-DEFENSE CORPS

APRIL 1963

Problems

There are too many different types of military and paramilitary organizations in this sector. The result is too many chains of command and too much uncoordinated activity. This is especially true in the intelligence field. The flow of information upward is confused, the flow of information laterally is worse, and the flow of information back downward leaves something to be desired. It is my experience that only the advisory effort is holding the whole thing together.

The Civil Guard battalion has become too large for itself. The original TO&E [Table of Organization and Equipment][3] calls for only five

---

3. In our army, TO&E was a list of all personnel skills and equipment in a given type of unit. We spent considerable time at the infantry school and at NATO in Germany studying these lists so that we all had a standardized understanding and could effectively

combat companies. This "battalion" now has twelve in various states of development. However, the Headquarters and Headquarters Company and the Administrative and Direct Support Logistics Company have not been correspondingly increased to support this additional strength. Besides this, these units must also support twelve SDC [Self-Defense Corps] platoons, twenty-one SDC squads, and an SDC training center. To further compound these organizational structural difficulties, the unit lacks officers, NCOs [non-commissioned officers], and vital communications and transportation equipment to support the basic TO&E, let alone the additional requirements. I further believe that the unit possesses inadequate indirect fire support, the heaviest organic weapon being the 60mm mortar.

Basic Proposal

The Civil Guard battalion should be redesignated a brigade, regiment, or battle group. The use of the term "battalion" in connection with these much larger units is adversely affecting the thinking and concepts of everyone who understands the term.

The armor platoon of the Civil Guard battalion should either be re-equipped with armored cars that can be supported and maintained or the organization should be dropped. *(I think they were old British armored cars from Malaya.)*

The Civil Guard battalion or "battle group" should have an 81mm mortar platoon of six tubes added to the TO&E to give it some indirect fire punch. *(All fighting was done with small arms and machine guns.)*

All SDC troops should be organized into Civil Guard companies and assimilated by the new brigade, regiment, or battle group.

[Editor's note: What originally followed here were three proposals containing detailed charts for reorganizing the units. They were described as (1) What Kontum Sector Would Look Like with a Civil

---

deploy the units. For example, in Germany I worked a lot with various TO&Es of the Infantry Rifle Company, which consisted of 6 officers and 190 men in three or four rifle platoons, depending, as well as a weapons platoon and a headquarters platoon. The TO&E listed the ranks and job skill requirements of all personnel, plus all the weapons, vehicles, and equipment. The army also had lists called the Tables of Authorization, which augmented the TO&E and provided certain additional resources for the area of operations you were in, such as NATO. I think the then-new anti-tank missiles came in this category.

Guard Battle Group and Two Battalion Areas and a Reserve Strike Force, (2) What Kontum Sector Would Look Like with a Civil Guard Battle Group or Brigade with Tailor-Made District Forces and a Reserve Strike Force, and (3) What Kontum Sector Would Look Like with a Flat, Purely Military Civil Guard Structure that Eliminated Districts from Military Control.]

These proposals were submitted at the end of my tour of duty. I never got a response.

ARE THE CURRENT CONCEPTS FOR EMPLOYING THE INFANTRY BATTALION IN OPERATIONS AGAINST IRREGULAR FORCES TACTICALLY SOUND? "YES, BUT..."

MID-1963

[Intro from the mid-1990s:] Upon return to the States I had a month's leave to reunite with my family and get set up at Ft. Benning for Advanced Officers Career Course ACAR #5. There were about 140 officers in the class. About 20 of us were just back from Vietnam, including Milt Craddock, now sporting gray hair, and Darryl Savage, who had earned a Silver Star and a Purple Heart with two oak leaf clusters. There were about 40 foreign officers, 20 of whom were Vietnamese.

We were taught the new battalion-brigade doctrine and shown all the new weapons we didn't have in combat. None of the classes, including the classes about guerilla warfare in the jungle, were taught by Vietnam veterans, mainly because there weren't many, as me and my fellow classmates were among the first to return home.

In one class, a captain was teaching us helicopter airborne assault of an infantry battalion against a guerilla force in the jungle. As the subject unfolded, we veterans started sneaking looks at each other as we watched our instructor unfold the action. Finally, Darryl Savage jumped to his feet and said, "Sir, Captain Savage. I'm sorry, sir, but it's my experience that that would be a very costly deployment." The instructor mumbled something about, "Well, it's the school solution." Then I jumped to my feet and said, "Sir! I'm afraid I have to agree with Captain Savage, sir. In the mountain highlands, that deployment would lead to a possible slaughter of our people on the LZ [landing zone]." Suddenly, the other recently returned combat vets started popping up around the room and shouting their assent.

With that, one of the school wheels, a big, burly bird colonel or a brigadier, I forget which, got up in anger and started down the aisle for the stage. I thought we were in trouble, because our army had an international reputation for its rigidity and predictability in combat. For being part of a democracy, our army was pretty authoritarian. The Germans in World War II allowed far more flexibility and give-and-take up and down the chain of command.

"All right, gentlemen, that will be about enough of that!" the big senior officer roared, red in the face and plenty mad. He walked back and forth on the platform and looked up and down at each of us who were standing. I think he wanted to say, "Who do you think you are? This is insubordination!" But instead he was man enough to respect our experience, despite our youth, and he said, "Each of you officers that is standing will write a white paper on some aspect of the war that you think you can teach us about. Each of you will be given a date to present your paper before the School. Now sit down, gentlemen!" I put together and presented the following.

Recent experience as advisors to the Republic of Vietnam combat forces lead us to scrutinize our own infantry battalion and determine whether or not the current concepts for employing the infantry battalion in operations against irregular forces are tactically sound. The scope of counterinsurgency operations is so great that any short discussion must be relatively general and incomplete, but certain areas are worth examination.

Area Organization
A vital consideration of area organization is to avoid assigning the battalion too much territory. Further, the area concept cannot be approached with the idea of uniform, stereotype organization. Each area and each sector will be different from others. The situation in each sector will be constantly changing, and allowances must be continually made for

this. Many factors must be considered when assigning battalions to sub-sectors. Not the least of these is the personality of the battalion commander and the state of training of the troops.

Commanders cannot be hard-nosed and prejudiced in approaching this mission, as there is no place for this type of mentality in counter-insurgency operations. The proper mental attitude on the part of the commander and his people requires tolerance for:

1.   Other peoples.
2.   Other walks of life.
3.   The necessity for skills other than purely military in these types of operations.

A possible type of area organization for an infantry battalion in places where companies are required to use reaction force and to clear sectors may have local friendly forces integrated and a battalion strike force retained. Battalion headquarters might be located with a battalion of local friendly forces and with a rifle company in reserve with an independent local friendly-force rifle company. In effect, this would comprise a brigade minus, a powerful strike force, especially if there was available helicopter lift capability.

A rifle company and a platoon of local friendly forces (LFF) might be located at a supportable distance, and the third rifle company and a platoon of LFF similarly located elsewhere. Various mixes of LFF companies and platoons might be located around the sector at key locations. Good communications and mobility are essential. The infantry battalion would have an offensive tactical role and would be aggressive in its patrolling. Careful planning of time/distance factors for various types of mobility, air assault, and vehicle or foot marching would be essential so that commanders not only have a good feel for when they can get troops where they need to be, but they can anticipate where and when the enemy can get in different scenarios.

Combat Bases

The present doctrine *(circa 1963)* calls for the use of combat bases. What are some of the considerations for the employment of combat bases?

If the companies are required to maintain a platoon(+) or (-) as a local reaction force in addition to conducting company clearing operations, the company combat base will usually have to be a semi-permanent fortified position. This will be dictated by the fact that the reaction force will have to be centrally located, where it can provide relief to those places the company mission calls for. Such a place may be located with roads or trails available for communications with key places under the company's protection. The combat base itself may double as a security post for a key place or installation. This type of base is not merely a static defense position. It is only that among other more important things. From here the maneuver platoons of the company will operate continuously in the company's assigned area. The company headquarters' organic and attached supporting weapons can be located with good security at the combat base.[4] Mortars and attached artillery (if any) can support operations and provide protection for key places within range. Possible enemy ambush along the roads and trails can be minimized by the aggressive actions of the maneuver platoons.

If companies are not required to maintain a reaction force and may devote themselves to "hunter-killer" activities or clearing operations against enemy guerillas, combat bases should become mobile. Secrecy and frequent relocation become characteristics of this type of combat base. The unit establishes and secures its base, clears the surrounding area, and moves on to repeat the procedure elsewhere.

Airmobile Operations

Supremacy of airmobile operations depends to a great degree upon the nature of the terrain and the type of aircraft used. A great many restrictions are placed upon the effective use of helicopters in guerilla terrain. Swamps, tall forests, thick jungles, and mountains have taken their toll on aircraft and personnel in these operations in the Republic of Vietnam. Effective enemy ground fire constitutes an even greater threat. Better and more transport and combat helicopters are needed, or else some superior machine. Losses of men and machines in these operations can be staggering. While airmobile operations can be highly effective, they have not been as effective as commonly felt. With present equipment *(circa 1963)*, the possibility of heavy losses must be considered in planning. Wise enemy guerilla commanders can, in certain situations, outmaneuver helicopter-borne forces. Helicopter-borne forces are *extremely vulnerable and can be ambushed, deliberately or by coincidence.* This must be

---

4. "Organic," as in the phrase "organic weapons," is an American military term meaning "permanently assigned."

considered in planning; set patterns must be avoided, and suppressive artillery fires and/or tactical air should support landings. *(Apparently no one paid any attention to any of this, because later US troops were inserted by helicopters into mutual surprise slaughters.)*

Task Organization

For a counterinsurgency mission, the battalion will require special tailoring based upon the particular situation. In addition to augmentation by intelligence, civil affairs, communication, military police, and other functions, deletion of certain elements may be necessary as an economy-of-force measure. As an example, anti-tank and Davy Crockett *(a tactical nuclear capability circa 1963)* elements of the battalion may be removed; or, these personnel may be trained in advance to assume certain counter-insurgency roles, thus minimizing the need for outside augmentation. If routes of communication exist in the battalion sector, the reconnaissance platoon can provide an excellent security force. Its functions may include road security, convoy security, and even reaction-force operations in relief of enemy attacks along routes of communication.

Communications

Effective communications is key to success in counterinsurgency operations. The battalions' communications potential may be augmented in a variety of ways depending upon the situation. The battalion commander, immediately upon entering his sector of responsibility, must centralize all existing communications under his control. This is of vital importance to the intelligence effort if information is to become intelligence and be disseminated to the proper people in time for action.

Artillery Support

When properly used, artillery has proven a decisive weapon against guerillas in the Republic of Vietnam. An excellent combat-tested method calls for decentralization of artillery to the point where even single weapons are employed *with* rifle companies. Guns are located within the security of combat bases. The optimum situation would enable interlocking fire at least in the most threatened portions of the sector. Some firing platoons might be employed in such a manner that they displace (with infantry security) *on alert* to support areas not normally covered. Fire direction elements can be divided among the weapons as required, and airborne FDCs (Fire Direction Centers) may be used. It is even possible to fire effectively using map data from relatively poor maps.

Engineer Support

Among the myriad functions the engineers can perform, the most important is the construction of roads and routes of communications. These provide increasing mobility to the infantry and aid in gaining control over the sector. New roads gradually restrict the previously larger area the guerillas could operate in. Roads can also provide a psychological impact upon the local people by showing them the extent to which friendly forces will go to protect them from the enemy. Furthermore, the region benefits economically, as a whole new vista of trade and markets are opened for the people. Engineers can change the political and economic geography of a large region as well as provide new routes for troops.

Ground Mobility

The concept of the infantry battalion being largely restricted to foot operations in guerilla warfare is not necessarily true. In certain situations, vehicular transportation may be the rule rather than the exception. The longer friendly forces operate in an area, the more the routes of communication will improve. Transportation will have to be pooled at some level, which will depend upon the time/distance and the situation. Thus it may be at the division, brigade, or even battalion level. Rifle companies will probably retain some organic transportation while reverting some to the higher-echelon pool of vehicles. Battalions will probably revert a lower percentage of vehicles to the pool. In certain situations, battalions may have sizeable transportation elements temporarily attached.

In the above, I have shared very briefly some thoughts concerning areas of interest in our counterinsurgency doctrine. These thoughts are based upon personal study, observations, and participation as a MAAG detachment commander and advisor in the Republic of Vietnam in 1962–63. The above topics concern those areas where I feel my ideas either differ from the current concepts as I understand them or where I feel more emphasis is necessary.

CHAPTER EIGHT

LATER WRITINGS

SOME PROBLEMS AS I SAW THEM

LATE 1990s

The advisory effort in Vietnam happened with a very fast ramp-up, but while World War II started up fast and smart, Vietnam was a mess. A lot of good information was being accumulated in the field, but it did not appear like there was any kind of feedback system for us to make suggestions. The top leaders made junket trips out to the field periodically, but they really didn't hobnob with the troops and get the picture. The histories of the period point this out pretty well. One really wonders how seriously anybody except us advisors took our efforts.

There was no special training in the area and culture we were going to. I was the only officer I ever heard of who had spent quality time in the months and years prior researching these subjects on my own. I also wouldn't have been as successful if I didn't have the background knowledge I'd attained before joining the service. During the year I spent in Vietnam, they used me as detachment commander for eight months and as a combat advisor for four months. If I hadn't taken university business and accounting courses at night school on my own time during my previous stateside tour, I could not have performed as detachment commander. It seemed as if I was the only officer with those qualifications around for a while.

There was also a lack of top-level strategic thinking when it came to how to fight in a guerilla war compared to a conventional war. By late 1960, as a senior first lieutenant, I was probably one of the few US soldiers who had researched counter-guerilla warfare to the extent I had. Given the fact that I came highly qualified and with superb tactical experience from Europe, I should have been a prime resource for the moment. However, it appeared that top-heavy bureaucracy in the army had already deteriorated to the point that the system didn't want to find people like me. They seemed to have no way to poll the officer corps to identify this kind of hidden talent, and there seemed to be a prevailing attitude that if you were not a West Pointer, you probably weren't worth knowing about anyway. Even though I found out in later years that I was apparently part of a hand-picked group, at that time no one ever seemed to know or care that I had this special knowledge, not to mention my proficiency in the Vietnamese language. The powers that were running the country and the war did not seem interested in finding such skills in people outside of their own group.

The language problem was also not handled well. The program consisted of sending some of us to an abbreviated language course before we went over. I was trained in Vietnamese in a two-month-long crash course with fifty-eight other officers, with the result that probably three of us had a good base to start from—that is, I could get around somewhat in the language. The training needed to be longer, or possibly augmented when we got to Saigon. The Viets furnished interpreters who did better with English than we did with Vietnamese for the most part, but they didn't know English as well as most Americans thought they did, and too much reliance was put on them. A lot of things didn't get translated properly, I would imagine, in those early days.

Former prime minister Nguyen Cao Ky provides an interesting account of the Vietnamese viewpoint of us; they thought we were rather patronizing.[1] For me, this was the result of several factors. One, few of us were able to "get into" the Vietnamese culture in the short time we had there. Furthermore, no one was listening to us. The Vietnamese quickly

---

1. Nguyen Cao Ky, *Twenty Years and Twenty Days: How and Why the United States Lost Its First War with China and the Soviet Union* (1976).

saw that our own top leadership was not only not paying attention to its troops, but on occasion they were also telling us to shut up and be good little boys. They knew in a short time we'd be going home anyway. A few years ago, I heard there had been something like six thousand books written on the Vietnam War. I wonder if there were any recorded incidents where senior officers listened to their young officers about what was going on *and then actually heeded that input.*

There was also a security problem in this connection that was probably worse than we suspected. After the war, we found out a lot of Communist spies had infiltrated everywhere. That had to be why the enemy so many times seemed to be "reading our mail." You couldn't put together an operation like the one I would have wanted to do for Tchepone [Xépôn] and keep it secret, let alone clear it through all the channels. There were so many chains of command that you couldn't propose anything without everyone getting involved, and such initiatives couldn't happen from the bottom up, they had to work from the top down.

Advisors also should have been better prepared and sent in for longer tours in larger numbers. I can't imagine having to be away from my lovely bride for any longer than I was, but if we were going to do it right, I should have been just starting out when I was already coming home. The original length of tour was eighteen months, cut short to twelve early in my tour. A one-year tour was not enough for any officer to make an impact. What was needed was a two- or three-year tour, with a chance to be with your family for a short time somewhere in the middle of it. Other officers should have been brought in to help, even in command roles, and we frankly should have built the infrastructure for the Vietnamese army to really train their junior officers to take over after us. (I say that my tour should have been longer, but the truth is that once I saw that the top leaders in the US government were not doing their due diligence and using "reasonable and prudent" management, I started counting the days before I could go home.)

One Civil Guard company at a time should have been in serious training and re-equipping. One way to have done it would have been to beef up the advisory effort by having a company officers' school, an NCO academy, and individual, squad, platoon, and company training at some fairly secure site. The companies would be rotated through this training one by one. At the time they went through this, the companies would be built up to strength as much as possible. I believe the Germans used a system like this in World War II, when units beat up in Russia would be pulled back and given time to reconstitute themselves. The American army, by contrast, had a terrible policy of letting units get chewed up in combat and then they'd ship in green replacements, who would be in grave danger from minute one because the unit was all shot up and the new guys couldn't be integrated intelligently. Things like this that happened in the twentieth-century US military just didn't jive with the dignity and rights of the individual that were supposedly guaranteed to us by the United States Constitution. There were a lot of those nasty little secrets that got swept under the rug from World War II on.

As I continually related in my letters home, the Viets were way short on officers and NCOs. Most of those they had were not competent to really train their people. It was hard for them to get better, although, as we saw, it was happening. No one on our side really seems to have looked at that and jumped on it. The whole war was handled throughout with a plethora of bad, bad leadership. It just doesn't seem like we could be the generation who followed all the sharp guys that fought World War II.

THE BEST AND THE BRIGHTEST?

LATE 1990s

When I got back to the Advanced Officers Career Course at Ft. Benning, I ran into Captain Darryl Savage, one of my OCS classmates. He was one of those guys I kept turning up with, like Milt Craddock. Darryl and I served at OCS, in line outfits in Germany, and at the army's language school, and we went into Vietnam together, although we were sent to different assignments. Now again at Ft. Benning, I found he had won the Silver Star and three Purple Hearts, along with the Combat Infantry Badge. I was dismayed that in a war like Vietnam they would send a combat soldier back into combat after his second wound. I was also dismayed that some of us had access to combat decorations that others didn't; some of us were told we "weren't in a war." There were many ways in which we were not playing on a level playing field, but in that case at least, a fine soldier received some recognition he deserved.

There were 140 officers in our class of captains and above. About 40 were foreign officers from other countries. I got an average grade on the first test and top grades on the rest for the six months the course lasted. I graduated in the top 10 percent and received a letter inviting me to apply for future admission to the Command and General Staff College at Ft. Leavenworth, Kansas. We knew in those days that this invitation meant you could "get your ticket stamped" and have an opportunity to make general.

But allegedly, even though I was selected in late '61 as one of what David Halberstam called "the best and brightest,"[2] I and many of my comrades found that daring to appeal policy, even through standard channels within the military, cost us our army careers. The "establishment" chose to stomp us out, contain us, or ignore us rather than listen to us. Some young officers resigned, most senior officers submitted. Later comers in Vietnam, such as Colin Powell and "Stormin' Normin" Schwarzkopf, cashed in (probably unknowingly) on the sacrifice of the "first wave," but thank God for the nation that these men were also quality soldiers and stayed the course to rebuild the army. We (the earlier group) failed to mount any reasonable attempt to bring sanity to the war, because we had no support from anywhere.

Andrew F. Krepinevich wrote in 1986 that no group of officers worked harder to change the army's approach to counterinsurgency than the advisors.[3] We were on the cutting edge of the conflict, receiving daily input through our own experiences on how well the operations being conducted were working. The reader can see that in evidence in the letters I wrote home to my wife and mother (chapters 2 through 5 of this book). Major Krepinevich, an officer of the next generation, continues to say in his excellent review of events that many of us advisors, and perhaps some Special Forces, were deeply convinced that the army was failing in South Vietnam. Krepinevich points out that, in our revolt from below, we advisors managed to disturb the government to a small degree, particularly during our heyday between 1962 and 1964, but once some of us saw that not following the party line could jeopardize their careers, advisors were more likely to just tell their superiors what they wanted to hear. This was the intellectual dishonesty. Officers were forced to make moral decisions they should never have had to make. It is no wonder the moral climate of the war began its descent.

---

2. David Halberstam, *The Best and the Brightest* (1972).

3. Andrew F. Krepinevich, Jr., *The Army and Vietnam* (1986), 80.

I was surprised to find in Krepinevich's book a statement to the effect that it was remarkable that a number of senior advisors in Vietnam—colonels and lieutenant colonels with a good chance of being promoted to general—jousted with the government leadership over the conduct of the war. After General Harkins made a report in March 1963 that said the military phase of the war could be won in 1963, several field grade officers spoke up to provide strong contrary evidence. Among those named were Col. Wilbur Wilson (I believe this was "Coal Bin Willie," at II Corps, later III Corps, and the boss of killed-in-action Lt. Col. Tencza and his successor Lt. Col. Sweet), Col. Porter (III and IV Corps), Lt. Col. John Paul Vann (7th ARVN Division), Lt. Col. Fred Ladd (21st ARVN Division), and Lt. Col. Rowland H. Renwanz (deputy of II Corps).

But the very nature of the military life at the officer level limited the ways in which an officer could appeal policy. Field grade officers didn't go around griping to young captains. Captains were limited in their ability to unload their views on superiors, depending upon the individual relationships involved. So we were all more or less isolated and individually suppressed or shunted aside. Given the system, we could not collectively launch a revolt; that's not the way the United States Army works. So, to use a slang phrase, "we got stuffed." This was not hidden to our top government leaders, and it should have been their responsibility to heed us and intelligently act on our information and recommendations. We did our duty, and each of us paid his own price for it. The government took advantage of the military hierarchy system and the professional integrity of appealing officers to divide and conquer amongst us and keep us silent. The government concealed from the public the attitude of the advisors, accused the opposition party of being warmongers, and then ramped up the war a few months later, after the 1964 election, doing things they accused the opposition of doing. Handling it the way they did was, I believe, a crime perpetrated against the people of the United States, especially the troops and their families, who became the direct victims.[4] Hiding all the records behind a cloak of secrecy for thirty years, together

---

4. In *Where the Domino Fell: America and Vietnam, 1945–1990* (1991, p. 82), James S. Olson and Randy Roberts recounted an interaction that happened during a 1967 briefing involving the Johnson administration's security advisor Walt Rostow and retired Lt. Col. John Paul Vann, who was a combat advisor in 1963 and was in '67 heading up the army's pacification program. Rostow asked Vann if he agreed that the war would be over in six months. Vann, laughing, replied, "Oh hell no, Mr. Rostow. I'm a born optimist. I think we can hold out longer than that."

with public apathy, leaves the whole thing in the hands of the Divine Tribunal. In short, I don't think there is anything we can do about it now.

———————

I have often wondered how it came to be that we early advisors and our charges were equipped with such obsolete weapons and equipment. I wondered if it was a deliberate move to get rid of the old inventory. If so, they also succeeded in killing off a bunch of us and getting rid of some people inventory too. It seemed a very strange thing for a wealthy superpower like the United States to send in its supposedly "best and brightest" with such obsolete weapons. The logic behind that follows most else.

In a 1998 documentary on the History Channel, several of the persons then involved blamed the weapons delays on a combination of the army bureaucracy and McNamara's "whiz kids," who kept interfering. Together, they did a good job of greatly limiting us in combat and probably indirectly killing a bunch of those whose names are now on The Wall. Personally, I believe "the best and the brightest" should have had the best stuff, since we were the ones doing the fighting in what was essentially an infantry war. Modern infantry combat is all about firepower, and the potential for return on investment in small arms systems was very high at the time, especially compared to the high-tech stuff the Department of Defense was spending on. I got slam-dunked for trying to make that argument when I started trying to make my appeals through my congressman in '64 or '65. Also, we now know that over two million US infantry weapons were left behind in Vietnam and Cambodia in 1975, and about half were state-of-the-art M-16 rifles, M-60 machine guns, and grenade launchers.[5] So, the facts of history are that we denied our few thousand "best and brightest" the best weapons we had at the beginning, and then we abandoned millions of weapons to the enemy at the end, including a million of our best weapons, things *we* could have had.

———————

I was a dedicated professional soldier, and the "establishment" smashed my career, as well as the careers of thousands of other good officers, and

———

5. Edward C. Ezell, *Personal Firepower: The Illustrated History of the Vietnam War* (1988), 155.

went on to kill fifty-eight thousand of our countrymen. The tragedy of history is that those who do these things seem to be helped by the world system to escape justice.

Nguyen Cao Ky, former prime minister of Vietnam, wrote, "Had the Americans arrived in our country with clear cut ideas, we might have learned. However, they arrived full of good intentions but without any real understanding of the problems involved, without any real policy; and so took refuge in makeshift accommodations."[6] If you use the analogy of a football game, the army kept running new players in and out of the game for the duration of the war instead of finding the winning combination and going with that, as in previous wars. Most of the players were prepared by the system, and that's what they had to offer. Some were shuffled into prestigious situations to "get their tickets stamped" for career purposes. There were also stories of some Saigon commandos who finagled situations to get themselves CIBs and other decorations. There may have been some others besides myself who spent years of preparing for this conflict in our free time. We went into the game for our short stint, and the system didn't care about that. If anything, we were seen as annoying when we tried to speak up. There was only so much we could do. For me personally, soldiering through in Kontum, I made myself familiar with the contents of my survival kit, studied map routes to the sea, and made a personal suicide pact not to become a prisoner of the Communists but to go down fighting, if necessary.

The problem in the earlier advisory effort was that we didn't stay long enough to really help. Just when you are getting a handle on the situation and getting acclimated to the culture and the language, BOOM! You are gone. Glad to be gone, yes. A smart way to staff the war effort? No. General Palmer said about advisors, "Their thankless, dangerous, and difficult work generally went unrecognized and unappreciated. All too often the US advisers were the unsung heroes in Vietnam, while American fighting units took the limelight and garnered most of the rewards."[7] I couldn't agree more.

You have to view the situation in the context of the whole century, especially the Cold War and our recent history, and realize that in spite of the enormous losses we sustained in the Vietnam War, we managed

---

6. Ky, *Twenty Years and Twenty Days*, 45.
7. Bruce Palmer, Jr., *The 25-Year War: America's Military Role in Vietnam* (1984), 54.

to emerge to a place of producing men and women of the caliber represented by Generals Powell and Schwarzkopf. Communism collapsed and without a Third World War. While the United States and its allies deserve some credit for this by holding the line all those years, I believe the real victory was The Lord's, and we better not accrue to ourselves the glory due to Him.

THE ANTIWAR MOVEMENT

LATE 1990s

The growing opposition seemed to me at the time to be treasonable in nature and not something military officers should have anything to do with. However, I read an article on the front page of a St. Louis newspaper in 1965 that reported that the government was dismayed that twenty-eight thousand army officers with between three and thirteen years of experience had resigned their commissions in recent months. (I was one of them, with almost ten years of active service, three National Guard and Reserve.) That was right at the time when the government was first committing US combat troops to Vietnam. The direct result was that those troops lost their experienced officers, and the kids went in with relatively inexperienced officers, which obviously must have greatly increased our casualties. What a tragedy! I read in the 1980s that around this same time, the Joint Chiefs of Staff, appalled at the war policy of the government, were contemplating resigning en masse. But unlike their more courageous juniors, they did not go through with it, hence the American people did not get the intended message from those of us in the informed officer corps.

One chapter of Harry G. Summers's *Historical Atlas of the Vietnam War* covers a brief history of this era.[8] A quote from that section reads, "Protestors saw the fact that 30,000 Americans fled to Canada as evidence of the bankruptcy of U.S. policy. But it was later revealed that more than 40,000 Canadians had come south to voluntarily serve in the

---

8. "The Antiwar Movement," in Harry G. Summers, Jr., *Historical Atlas of the Vietnam War* (1995).

U.S. military." If you calculate those numbers as a percent of the respective national populations, it looks vastly worse.

In another of his books, Summers writes a long essay on the antiwar movement, giving a good overview of the subject and looking at it from many different angles.[9] Summers shows that there is a very big difference between appealing policy through loyal opposition versus through the kind of opposition that erupted from various elements during the Vietnam War. What confused me was that our country's involvement in the war was escalated by the liberal-democratic Kennedy and Johnson administrations, yet some of the noisiest opposition was coming from people in that same camp.

The shabby nature of much of the antiwar movement ruined legitimate, intelligent attempts at opposition being made by qualified people, such as returning young officers who sacrificed their military careers to try to appeal policy. Our efforts were lost in history. As for me, when I saw the scope of the effort required, I soon realized I wouldn't be able to do it without a sponsor, and I had no luck in finding one. (I'd sent a letter to the Pentagon via my congressman, but my congressman didn't even read what I wrote.) It also soon became evident that I would probably be linked with the antiwar movement, which was a label impossible for me to accept. Some of those people were abusing their privileges of American citizenship to persecute soldiers, sometimes employing libel and slander. And the politicians, rather than step up and take the blame themselves, just let them get away with it. The politicians sort of sneaked out the back door and let the troops take the heat. The politicians also did a very good job of hiding the evidence until the nineties. Summers points out that of the 7,575,000 American service personnel who served during the Vietnam War era, there were only 24 cases within the combat zone when a soldier deserted to avoid hazardous duty. There were 32,000 total desertion cases related to Vietnam, and many more men went AWOL, but only 24 deserted in the face of combat.[10]

The media, however, seemed determined to present two types of citizens during the Vietnam War period: doves, who wanted the fighting to stop, and hawks, who wanted it to be more aggressive. There was no shade of gray; you were one or the other. By presenting things in this

---

9. Harry G. Summers, Jr., *Vietnam War Almanac* (1985), 80.

10. Summers, *Vietnam War Almanac*, 139. See also *Webster's New World Dictionary of the Vietnam War* (1999), 95, 102.

way, the reputation of the media, already much tarnished by the war, was tarnished still more. This forced representation of black hawk and white dove totally misrepresented the facts, and in perpetuating the dichotomy, the media caused a great deal of national dissension. The truth is, there were all kinds of doves and all kinds of hawks. Only a few were traitors; only a few were fascists of the military-industrial establishment. There were a lot of shades of gray. I couldn't peg myself as one or the other, the way they were doing it. I was a hawk officially, because I wanted to do what it took to save South Vietnam, but I had dove tendencies, because I had a "Let's win or get out" attitude. Remember, resigning my commission was 80 percent an attempt to appeal policy and 20 percent a move to save my own life for my family. I would have been sent back, and back again, I'm sure. Maybe with the 1st Air Cav in '65, as I was a hot item for that outfit, having already worked the concepts in combat on the ground where they were going.[11]

You could call us idealists, I suppose, but there were a number of Americans like myself who were concerned about Communist imperialism and expansion, and we thought the time had come to draw a line in the sand and say, "No further." As the war was pumped up, opposition seemed to come from the left, and to some of us that meant the pro-Communists—basically American citizens aiding and abetting the enemy. We had difficulty understanding that some of these people had higher moral reasons for opposing the war. Now I know there was a sound basis for the story of the other side, but at that time it was not being told well enough to be heard and understood from my perspective.

When I came back from Vietnam, I was among the first of many younger officers who sacrificed his career trying to stop what I saw coming. Our seniors, the generals, let the nation down when they did not back us up by also resigning. Without the visibility that their resignations would have given to our position, they effectively consigned our sacrifice

---

11. In late 1965, the army prepared a special division, the 1st Air Cav, equipped and trained for airmobile warfare with helicopters. It's possible if I had stayed in the army as a senior captain, or maybe a new major, I might have been put into this unit for a second assignment to Vietnam. In November of 1965, the 1st Air Cav was part of the Battle of Ia Drang Valley, in which the US lost 234 of 457 soldiers involved. The enemy had ten times the losses, and our side saw it as a victory. The high-ranking US decision-makers and General Westmoreland felt they still had a conventional war they could win, not realizing this was an early battle of the type that eventually ground our line divisions down in fighting strength and turned the war into a meat grinder.

to never-never land. I *wanted* to force the Communists to back down and leave us and our allies alone, but I did not want to fight a fight that was, even to a young captain in as early as 1962, plainly hopeless. I kept supporting the war in the hopes that the government would "wise up."

In the late 1960s, James Reston of the *New York Times* wrote, "The art of resigning on principle from positions close to the top of the American government has almost disappeared. Nobody quits now as Anthony Eden and Duff Cooper left Neville Chamberlain's cabinet, with a clear and detailed explanation of why they couldn't be identified with the policy any longer. . . . Most (of those who stayed), at the critical point of escalation, gave to the president the loyalty they owed to the country."[12]

Quoting Napoleon, military historian Jeffrey Record expressed a similar sentiment: "Every general is culpable who undertakes the execution of a plan which he considers faulty. It is his duty to represent his reasons, to insist upon a change of plan; in short, to give his resignation rather than allow himself to become the instrument of his army's ruin."[13] Although a student of Napoleon as a young officer, I was not aware of this statement during my service, but I inherently knew that this was the way to act, and I did so. I trust many of my peers also acted because they had similar thoughts. And yet with dismay we watched our seniors cop out on this one. In his book *The Wrong War*, Record presents a story of Vietnam military leadership in ways that corroborate the story I attempt to tell in this book based upon my own personal experience and research.

When I got home, I became quickly disgusted with the antiwar protesters, many of whom reeked of anarchy, chaos, lack of morals, high crimes, and treason. Jane Fonda's trip to Hanoi, for example, I viewed as high treason. In some cases, I suspected it was a front for cowardice. Having been involved in the war at the beginning, I endured many long years of watching the bad treatment of the soldiers and the type of news coverage that was happening.

From time to time, I was also personally abused. On one occasion in 1965, I was asked to speak about the war to a group of people gathered in a large, posh suburban church. Some in the group refused to believe some

---

12. Reston quoted in Paul Hendrickson, *The Living and the Dead: Robert McNamara and Five Lives of a Lost War* (1996), 297.

13. Record quoting Napoleon in Jeffrey Record, *The Wrong War: Why We Lost in Vietnam* (1998), 160–61. Napoleon quote from David G. Chandler, *The Military Maxims of Napoleon* (1988), 79.

of the things I said and called me a liar. *I* was there, *I* was qualified, *they* were not. They publicly accused me of lying in the church.

One evening at home I was watching the news about a large mob of protesters at Washington University in St. Louis. They were about the business of burning down an ROTC building, plus engaging in other vile behavior. I was hard pressed not to take my shotgun and go down there.

Another time, a business prospect blew me off one day, saying, "If you were part of the army over there you must be a loser." And years later, in 1979, when I was a key expert witness in a corporate lawsuit, an attorney asked me not to let the jury know I was a Vietnam veteran, as it might jeopardize his case.

Through the months and years, I witnessed numerous digs made here and there, some while watching and reading the media, some said to me personally.[14] I was walking around with a wound that didn't seem to want to heal, and at one point I considered moving my family to Australia. These are just a few examples of what I and my fellow Vietnam veterans endured.

Among the positive interactions I've had were two or three separate occasions of someone coming up to me in church and saying they knew I was a veteran. After identifying themselves as former protesters, they proceeded to apologize for their part in the protest movement. Each of them said they were making these kinds of apologies to every Vietnam veteran they could. This was in the 1980s.

But as for the antiwar protesters at the time, there was a huge difference between the kind of protest most activists were making (including Bill Clinton, when he was a protester) and the kind of protest we soldiers were trying to make. We started within the system, working from the angle of discussion and policy appeal, and when that didn't work, we

---

14. American television quickly became an instrument of torture—perhaps unintentionally—for the returned veteran. Concerned about our comrades still in Vietnam, we watched with great interest what the media was saying about the war. We took a great deal of abuse by doing that, right in our own living rooms. For me it lasted continually from 1963 to 1975. You couldn't get away from it. After the war, it continued sporadically for me until Operation Desert Storm (1991) returned the army to its respected status in our society. For a period of sixteen years, from 1975 to 1991, I did my best to avoid the subject entirely, and it wasn't until 1995, when I read Robert McNamara's book *In Retrospect: The Tragedy and Lessons of Vietnam* that I decided to write this book.

resigned our commissions, destroyed our careers, and tried to continue the protest under rule of law. Among us commissioned officers, there is a point of disagreement about at which point it was no longer proper for us to do anything except resign—or knuckle under. No one in the press media, or in the establishment, to my knowledge, properly showed the public the range of views that existed. I haven't seen any evidence that this occurred, but I would like to be corrected by someone who could show me some true historical facts to the contrary.

Here is what a former enemy, Stanislav Lunev, a former colonel in the GRU (Soviet military intelligence), had to say about how the GRU influenced the American public: He said the GRU and KGB (the Soviet Union's internal security agency) helped to fund "just about every antiwar movement and organization in America and abroad. Funding was provided via undercover operatives or front organizations. These would fund another group that in turn would fund student organizations. The GRU also helped Vietnam organize its propaganda campaign as a whole. What will be a great surprise to the American people is that the GRU and KGB had a larger budget for antiwar propaganda in the United States than it did for economic and military support of the Vietnamese. The antiwar propaganda cost the GRU more than $1 billion, but as history shows, it was a hugely successful campaign and well worth the cost. The antiwar sentiment created an incredible momentum that greatly weakened the U.S. military."[15] I personally knew this was going on from the late 1950s, and it was a huge source of continued frustration for decades to be forced to observe how certain liberals were running the war while others were buying off on this propaganda and opposing it, and we couldn't get our house in order. And others like myself who were trying to appeal, "Hey! Hey! Come on! Wake up!" were swept into oblivion. Finally, the perpetrators classified the whole mess as secret and hid the facts for thirty-plus years. I can only assume it was to avoid accountability.

In the early nineties, Col. Harry G. Summers wrote this about antiwar sentiment: "The Vietnam protest movement generated negative feelings among the American public to an all but unprecedented degree. In a poll conducted by the University of Michigan in 1968, the public

---

15. Stanislav Lunev, with Ira Winkler, *Through the Eyes of the Enemy: Russia's Highest Ranking Military Defector Reveals Why Russia Is More Dangerous Than Ever* (1998), 78.

was asked to place various groups and personalities on a 100-point scale. Fully one-third of the respondents gave Vietnam War protesters a zero, the lowest possible rating. Opposition to the war came to be associated with violent disruption, stink bombs, desecration of the flag, and contempt for American values."[16]

In the aftermath of the war, mea culpas from former draft dodgers were commonplace, such as James Fallows's 1975 essay "What Did You Do in the Class War, Daddy?," which fatuously predicted a class war between those who fought overseas and those who used their privilege to evade that fate.[17] Fallows and his fellow antiwar activists misjudged their own importance, however, and within a decade the war protesters had fallen from their high moral plateau to the depths of derision. Addressing a student group at the State University of New York at Stony Brook in 1985, a former antiwar activist said that the reason he had opposed the war was to bring the boys home safely. A student then contemptuously asked, "When you spit at them, and called them baby-killers, and threw rocks at them, was that just your way of saying, 'Glad to see you'?"[18]

The only thing the antiwar protesters accomplished was their own selfish ends. They sabotaged protests trying to happen within the system and gravely damaged both the nation and themselves.

When you read the personal letters included in this book, you can see that we were helping the people of Kontum Province defend their homes from predominantly outside aggressors who committed many crimes against the civilian population. Knowing that, you can imagine how the antiwar protesters back home looked to someone like myself, who knew the war from the perspective that I did. We knew those protesters at best as ignorant fools and at worst as downright traitors. And our government looked bad too, putting us into a scenario like that and leaving us holding the bag.

---

16. Harry G. Summers, Jr., *On Strategy: A Critical Analysis of the Vietnam War* (1982), 14–15. See also "The Antiwar Movement," in Summers, *Historical Atlas of the Vietnam War*, 158.

17. James Fallows, "What Did You Do in the Class War, Daddy?," *Washington Monthly* (1975).

18. For a collection of interviews with Americans who publicly opposed the Vietnam War and who traveled to Hanoi to demonstrate their commitment, see James W. Clinton, *The Loyal Opposition Americans in North Vietnam, 1965–1972* (1995).

EPILOGUE

Through all the years, the top national leaders of this country never seem to have come to grips with the fact that Ho Chi Minh and General Giap and their team whipped and outwitted President Kennedy and Secretary McNamara, Generals Taylor and Harkins, and our team. They continued to whip and outwit the Johnson administration. The leaders on our side seemed content to let the antiwar protesters and, yes, the press blame the troops. They used us, the troops, in several ways. By laying the blame on us, they were able to let the fog of time make it impossible to bring the guilty ones to justice. I would put some of them up there with the famous war criminals of history, and these guys got away with it.

The key thing about Vietnam is that a relatively small amount of intelligently directed decision-making at the critical moment (1961–63) would have headed off the disaster that followed. In the early years, the opposition was greatly impressed with us and fearful of our warmaking capacity, based on what we had demonstrated in World War II, and I believe they were extremely leery about provoking us. The whole thing came down to timing and place. If we were not going to take care of things in Laos from mid-1961 to early 1963, we needed to just stay away. As it turned out, our enemies must have been pleasantly surprised to discover they were up against a stupid, arrogant establishment supported by a military bureaucracy of sycophants who were so impressed with themselves that they couldn't see what was going on.

This war was a classic violation of the Principle of Economy of Force. Because of the leadership that we had at the top, America sacrificed 58,159 men killed, 304,000 wounded, and 74,000 maimed. I have not seen any figures that agree on how many are still in veterans' hospitals. We also do not know how many veterans and their families suffered mental and financial damage, some of whom have not emerged from those conditions. One source reported that more than 43 million Americans lost someone in Vietnam, and that more than 100,000 veterans have died prematurely since the war.[1] I do know in my own little sphere of life in the Heartland that numbers of veterans with problems are still coming to my attention in the late nineties, and for me, incidents of personal abuse did not stop until the mid-eighties, twenty years after I came home.

As for the others, 224,000 South Vietnamese troops were killed, and more than 1 million were wounded.[2] Over 300,000 South Vietnamese civilians were killed. The enemy claimed 1.1 million soldiers died for Hanoi. North Vietnam lost about 3 percent of its population to battle, a level nearly unprecedented in the history of warfare, but they stayed the course.[3]

The Montagnards, my comrades-in-arms in the 10th Civil Guard, were almost taken out as a people, losing 200,000 killed in the Central Highlands, where I was, and 600,000 refugees. Over 2 million Asian lives are said to have been lost, and how many disabled? Vietnamese losses are probably beyond reckoning. We spent no less than $220 billion, and on the other side the Soviet contribution may have been a huge factor in the ultimate collapse of the USSR. Vietnam itself was impoverished for two generations. It is said 10 million Americans were air-lifted to Vietnam by commercial aircraft, more than 5,000 helicopters were lost, and 6.5 million tons of bombs were dropped. As for the number of lives changed or ruined in other ways . . . who can count?

And who was held accountable for this? The people who did this awarded themselves medals (e.g., the Medal of Freedom given to Robert

---

1. Paul Hendrickson, *The Living and the Dead: Robert McNamara and Five Lives of a Lost War* (1996), 124.

2. *Webster's New World Dictionary of the Vietnam War* (1999), 58.

3. *Webster's New World Dictionary of the Vietnam War*, 30.

McNamara in 1968), hid the records in top-secret archives for decades, continued to provide themselves jobs in the political system, and let the troops and the next administration take the blame. Can it be that the rest of us deserved what we got because we let them get away with it? Could anyone have run a more expensive war in terms of national treasure than these two opposing leaderships did? Hitler and Stalin did it in World War II. Both have been branded war criminals. Meanwhile, the leaders on both sides in this war got away with it, at least for now.

One thing I haven't written much about in this book and yet consider an important part of my life is faith. I was raised to the biblical worldview, meaning my fundamental values came from the Bible, and I perceived my professional military career through that lens. To me it was a vocation requiring a high sense of ethics, and you might even say I had a Divine calling to become a soldier. I believed the US was obligated to use its military legally and also righteously. I had a Crusader mentality—perhaps a bit naive and immature—and I was something of an idealist. My focus was Duty, Honor, and Country, and the army itself taught us this was the mindset we should have.

In Vietnam, in the region and time where I was, it was obvious that one of our tasks was to protect Christianity in our area of responsibility. This included the Church and its people and property, including French and American missions and hospitals that were there. These organizations were usually the only places offering medical services, taking care of orphans and lepers, and providing schooling. These Christian (predominantly Catholic) institutions were one little glimmer of hope while the South Vietnamese government was struggling to get established. Many of us advisors believed that righteousness and justice were our cause, and I think the people of Vietnam saw us as valuable protectors of their rights.

In reading literature about the Vietnam War, one often sees that the motives and value systems of the people at the top of our government apparently differed greatly from the motives and value systems of many of us in the fight. Realizing this caused a crisis in my life. The leaders of the system that built me into a man and a military leader were hypocritically violating the very values they had taught us, and yet when we questioned it we felt threatened and swept under the carpet. By the mid-1960s, I was reeling through life with a shattered value system and a sense of outrage. I am a strong person, but even I could not go on like that. How was I to support my family? I saw God and religion as a part of the problems I was suffering, and I eventually put Him on hold too. I started to shift toward the modern worldview that everybody else seemed to be following: Look out for number one. But I simply could not connect the values I was brought up believing in with this new liberal, modern-worldview lifestyle.

In 1967 my wife and her father encouraged me to re-evaluate Christianity, and I soon had a renewed personal experience with God, becoming "born again." I have often wondered what it would have been like in Vietnam with *this* kind of Christian foundation.

My family has served this country in war from the Civil War to the Vietnam War. I went into Vietnam as a young captain trusting the government to use the army wisely. But now, and for good reason, I do not trust the lives of my sons and grandsons to the government and to the military. It is past time that we redefine how America makes war.

Making war is the ultimate right of the American people. The politicians and the military are the servants of the people. In the early Vietnam War, it worked the other way around, and that was a big part of the problem. In his *Vietnam War Almanac*, Col. Summers discusses morality and war based on concepts derived from the fourth-century writings of Saint Augustine, as refined by Saint Thomas Aquinas in the thirteenth century.[4] The theory of a "just war" has two parts: *jus ad bellum* (the decision to go to war) and *jus in bello* (the conduct of war). It sets three requirements for the decision to go to war: the decision must be made by a proper authority, it must involve a just cause, and it must have the right intentions. For the conduct of war, it also established three requirements: the means used must be in proportion to the ends to be achieved, discrimination must be used to avoid harming non-belligerents, and civilians and the means used must be in accordance with the positive laws of war. From a legal point of view, morality is first defined in terms of the positive laws of war, primarily according to the Hague Convention of 1907 and the Geneva Convention of 1949. In addition to these

---

4. Harry G. Summers, Jr., *Vietnam War Almanac*, 255.

treaties, to which the United States is signatory, within the US military the authority to "make rules for the government and regulation of the land and sea forces" is delegated by the Constitution to Congress.

Summers points out that this practice changed in 1950, when President Truman committed American forces to combat in Korea without a declaration of war, and with this precedent, Presidents Kennedy and Johnson committed US forces to combat in Vietnam. Similar actions have continued to happen in the second half of the twentieth century. There remains controversy over who has the power to wage war, and how it shall be done, but no solution is forthcoming. And yet this is far too important a loose end in our system of government. We can't wait to figure it out until a crisis is upon us; we need to resolve this in peacetime, before we are put to the test.

How we make war should be a well-defined set of procedures for every potential enemy (and every American citizen) to know. We have been too careless about the lives of our citizens and soldiers in the last couple of generations.

In his conclusion to the fine book *The Year of the Hare: America in Vietnam, January 25, 1963–February 15, 1964* (1997), Francis X. Winters writes:

> The lesson of Vietnam is now increasingly inescapable: the fate that befell the American intervention in Vietnam was the ever-bitter fruit of colonialism. For the self-complacent American rush to remake Vietnam's government in a Western democratic image was a blind violation of the prerogative of sovereignty, the right of self-determination. Less abstractly, Kennedy's coup vainly sought to extinguish in Vietnam the flame of freedom in the name of an alien political ideal, democracy. Finally Kennedy's place in history will always be shadowed by his own, perhaps unconscious, colonialism, which mirrored the colonialist ethos of the correspondents and editors of his time.

As a young officer, a captain of paratroopers, a commander, and then an advisor, I now realize that the historical evidence proves we were committed under false pretenses. We were as colonialist as the French. When we questioned and appealed policy, we were ignored or told to shut up. Now, even the enemy says we were right, even in the face of Mr. McNamara, one of the perpetrators. No one has been called to account for this, or probably will be, and, considering the cost to all parties, it is one of the greatest crimes in American history. Isn't anyone going to do something about this? Shouldn't we? I know who I would like to see answer these questions, and I wish they could be forced to do it while standing in front of The Wall.

THE PRINCIPLES OF WAR IN 1961

Between 1959 and 1962, I compiled a paper that aimed to show how gue-rilla warfare should be conducted using a modified version of the Principles of War formulated by the Prussian general and military theorist Carl von Clausewitz in 1812. I submitted my piece through the appropriate army channels for possible publication, and I even had hopes of interest from Reader's Digest *and one other major publication. However, even though the media was ever more focused on the subject, my articles didn't attract attention, possibly because they were too technical, maybe written too much like an army training manual. In any event, they were rejected by* Mili-tary Review *and* ARMY, *with the paperwork reaching me in Vietnam in August 1962.*

In 1982, Colonel Harry G. Summers, Jr., published his On Strategy, *which used the perspectives of the Principles of War to cover what went wrong in Vietnam. The book became a key text in top army schools. Mean-while, my paper on the same subject had been submitted in 1962, hoping to offer some thoughts twenty years earlier, before we had fifty-eight thousand lost lives on The Wall. I submit my paper here as hard evidence that, even as early as 1961, we had the resources in our national inventory to do it right.[1]*

First, a bit of background: The Bay of Pigs fiasco in Cuba in April 1961 gave the Kennedy administration some concepts about guerilla warfare that later led to trouble in Vietnam. The media coverage of the successful Cubans, led by Che Guevara, seemed to spark the focus of the American civilian leadership and started them casting about for solutions. Their answer was the Special Forces. Actually, the Special Forces was made up of groups in different countries that were to be inserted behind the Iron Curtain as guerilla teams. Once the Americans were committed, Special Forces represented one of the most awesome threats out there. I could imagine the havoc they might have caused in Poland if they were, shall we say, properly used. But instead, the Kennedy administration turned them into an ad hoc counterinsurgency force. They really became a gimmick. There is no evidence that anyone deploying the Special Forces in Vietnam used tactics based on the Principles of War. This was their first war, and they came out one loss, no wins. Considering the quality of the manpower, they deserved better. The Kennedy administration's strategy of counterinsurgency failed, largely due to wholesale violation of the Principles of War. I address these issues in the article that follows.

THE ROLE OF THE STRATEGIC LEVELS IN IRREGULAR WARFARE

By Capt. M. J. Frankwicz, Airborne Infantry, USA, researched and compiled in 1959–1962 and submitted in early 1962, returned on 1 August 1962 cleared for publication but unplaced

The decisions in irregular warfare are gained or lost at the strategic level. This theory seeks to enhance the importance of the strategic-level decision in irregular warfare because the twentieth-century experience has revealed a need for such a reawakening, as it were. At the same

---

1. See also Franklin Mark Osanka, ed., *Modern Guerrilla Warfare* (1962). This book was published while I was in Vietnam, so I didn't come across it at the time. It is a compilation of articles written mainly by various military officers about Communist guerilla movements from 1941 to 1961.

time, however, we must not detract from the importance of tactical competence.

Conventional warfare can witness the most competent strategic-level plans go awry because of the tactical brilliance of an enemy general. This is because tactical forces in conventional warfare can be sizeable enough to gain a decision. The relative importance of the two is more nearly parallel. In fact, poor overall strategy (in comparison to the opponents') can be overcome by brilliant tactical successes of a few generals. The Axis demonstrated this to some degree in World War II.

But in irregular warfare, while tactical competence is highly important (indeed we do not seek to detract from the emphasis upon it), top-level strategy decisions, important in any war, are actually enhanced in irregular warfare. The very nature of the tactics depends too much on the multidimensional aspects of the strategic level. Indecision or the wrong decision from the top are not easily compensated for in a guerilla war at the tactical level. The key tactical commanders are captains, lieutenants, and sergeants. The chances of having hundreds of them capable of brilliant tactical maneuvers in a strategic situation which has stacked the odds against them are remote. But in conventional warfare, the emergence of a few brilliant generals can affect a tactical situation which will bear more heavily upon the outcome of the war.

Tactics of irregular forces in a guerilla war are small-unit tactics that generally begin on a rather crude scale. The overall strategy of the guerilla forces is definitely predictable. To be successful it must evolve through several stages of increasing strength. Strategy and tactics of forces opposing irregular resistance movements, usually conventional forces, have a tendency to vary their approach with each war. The definite pattern set by military history for the guerillas is a result of the recent repeated success of these guerilla methods. Success has been assured in most instances by the relative uncertainty of the strategy and the tactics of their adversaries. Even so, in recent years a pattern for anti-guerilla warfare has emerged. But it is a pattern that has largely concentrated on the tactical aspects of the struggle *after the fact*—after the irregular forces have emerged in some strength. The current *(circa 1961)* anti-guerilla warfare pattern has failed in some areas where it was employed because, in the words of the Germans who developed it against the Soviet Partisans, it was "too little, too late"; a strategic flop for the Nazis. The British applied the pattern successfully but at great cost in Malaya. It worked not merely, as some British officers suggest, because it was a jungle war of platoons and squads but principally because the situation was diagnosed soon enough at Britain's strategic levels. Reaction and competence at these levels was prompt enough to enable their application of the tactical pattern to be successful.

The Mental Approach

It is vital for the strategic level of any nation which could find itself involved in irregular warfare to *know* guerilla warfare. By this we mean the intimate diagnosis of this type of warfare, and the different applications of the Principles of War to this style of war, as opposed to the frame of reference conventional fighters are used to. The strategic level must be able to attune the mind to the guerilla warfare channel and resist the temptation of "mental jamming" from the reference manuals of conventional and nuclear warfare. And this will, of necessity, require as much forethought on national-planning levels as other war-deterrent factors. If our empirical observation leads us to believe that this thought is too ridiculously simple to merit our consideration, then we have good reason to develop our subject further. Our task here is to analyze why it is not only logical that we consider the obvious strategic role in irregular warfare, but vital. We shall attempt to show that it is a role that has often been abused in history, to the extent that it most certainly merits our consideration. We shall endeavor to analyze what this role must be from the military point of view. History has shown us three reactions by anti-resistance forces in the face of our tenet.

First, there has been the nation that was ignorant of its potential fate at the hands of the guerillas until too late. France fell victim to her English and Spanish enemies during the Peninsular Campaign, and also to the Russians at the end of that tragic campaign during the Napoleonic Wars. So also did the Turks falter against Lawrence of Arabia in World War I. Nazi Germany was also such a nation and paid in full for her ignorance over most of the territory she conquered during World War II. Perhaps we can also attach this label to the Japanese during the same time, for China, Southeast Asia, and many Pacific islands caught Japan's forces in the web of irregular warfare. Batista's Cuba was a more recent example of the first type of national mental reaction.

Second, we have the nation that foresaw the trend of events to some extent, but lost because they were unable or unwilling to pay the price.

The earliest example is the British Crown, which during the American Revolution lost the American colonies to this type of reasoning. The situation of the Chinese Nationalists seems to belong in this category as well. But unlike the British, who seemed unwilling to pay the price, the Chinese were both unwilling and unable to. The same would seem to apply more or less to the French wars in Indochina and Algeria. This also appears to be largely the case today with South Vietnam, Laos, and certain other South American countries and other world hotspots that have not erupted as yet. Most of the old colonial powers fit with this category too, although some fit in the first group.

Third, we have the nation that foresaw the problem, had the necessary resources, was willing to pay the price (although at times this meant withdrawal), and was prepared. In many cases, this type of nation was able to extricate itself without any armed conflict. The most successful example of this appears to be the current period in British history. In many cases, Britain's policies seem to have enabled her to move relatively calmly through the transitions of her colonies gaining their independence as nations. Violence flared from time to time, but proper resources were employed to restore order in most cases. Partially fitting into this category is the example of American policy over the years with the Philippines, which sometimes resulted in good relations. And we must not overlook the reactions of the Soviet Bloc toward the possibilities of "counterrevolutionary" struggle, which, while diametrically opposed in method to those we have discussed, represent alertness of the highest degree.

Strangely enough, the most useful *tactical* patterns for combating guerillas have been developed by nations in these first two groups. A German armed forces directive of 1944 laid down a tactical pattern for use against guerillas that formed the basis for much current *(circa 1961)* thinking and doctrine on the subject. French operations in Indochina against the Viet Minh also provide many contributions well worth study. It was not that these tactical doctrines failed, but rather they were developed too late. Remaining resources were too meager to win, however brilliant the tactical doctrine.

We have outlined the types of mental reactions of governments opposing resistance movements. There is something to be said for the mental approach of the resistance hierarchy as well; in a world of human weakness, the side which is aware of the weak spots of the enemy and of its *own* weak spots is the side which gains odds. For that is the side which can minimize the effects of its own weaknesses and prey upon those of the enemy. Recent history, as we have summarized, abounds with lessons we dare not overlook.

The resistance movement which approaches its forthcoming struggle with an inadequate frame of mind and lack of knowledge is doomed to a speedy failure. History is replete with accounts of abortive revolutions. Each failed primarily because of the inadequacies of its strategic level of leadership. Many of the street revolutions of nineteenth-century Europe are examples, as is the failure of the Communist strategy in Malaya after World War II.

There is also a type of resistance movement which suffers near fatal leadership pains at first but then slowly improves its leadership over time and to the degree that victory is possible. Usually predominant factors that enable a resistance to survive (besides luck) are weakness and even greater ignorance on the part of the enemy. Outside help from a sponsoring power can be another key factor in improving resistance leadership in this type of situation. An excellent case in point is the opening of the Viet Minh war against the French in Indochina. The Viet Minh all but blundered themselves out of existence in the beginning. But the weakness of the French forces, the mental attitude in Paris, and poor luck confounded the best French efforts. When the Red Chinese arrived at the border, the Viet Minh gained a sponsoring power. Among the decisive forms of assistance the Chinese provided was training of the Viet Minh leadership to a degree of skill sufficient to gain strategic superiority.

In the last category are resistance movements whose leadership is strategically competent from the outset of the struggle. The revolutionary movement blessed with such leaders will apply a strategy that can win, and usually does. Only the shrewdest opponent can defeat such a movement. The best example of such strategic leadership is the Communist camp and all the machinery used to produce it. The Communists train their potential revolutionary leaders well in advance. For this reason, they are capable of producing highly competent leadership at all levels from the very outset in their revolutionary wars. The prime pioneers of the "system" were Lenin and Mao Tse-tung. The most prolific was Mao, who provided a great variety of precedents for action against strong odds. The recent Cuban revolt of Castro and Guevara shows the potential product of this type of strategic leadership.

The importance of the relative relationships between the mental approaches of the resistance and anti-resistance camps is vital. Positive thinking, open-mindedness, and sound psychopolitico, economic, and military backgrounds in this manner of conflict at the top levels are prerequisites for success. Both sides must have these qualities before the military phase of the struggle begins. For the resistance movement that seizes the initiative, this is obvious; for the anti-resistance party, this is a much-violated rule that has often led to disaster. The anti-resistance group will find the struggle thrust undeniably upon it, and therefore it *must* be prepared.

Having examined various aspects of the mental approach, we shall now consider some of the military aspects of this form of conflict.

The Anatomy of Irregular Warfare
A diagnosis of the anatomy of irregular warfare will reveal that the conflict increases in scope by progressing through a definite number of clearly defined stages.

Stage 1. The existence, creation, or importation of conditions favorable to the birth of a resistance movement.

Stage 2. The incubation period. This is the beginning of a resistance movement, and it is characterized by passive objection, individual expression of opposition, minor sabotage, individual violent action, and organized group violence. By this time, the handwriting will be on the wall. As dissident groups become convinced that their cause will not be served by legal means, they become increasingly frustrated into violent eruptions.

Stage 3. The guerilla warfare buildup and preparation phase. At this point, the guerilla element of the resistance group begins to form, organize, and train. Bases are established, supplies are acquired and stored, and intelligence and communications nets are developed. Psychological warfare in all its aspects plays a predominant role as propaganda is manufactured and spread. The guerillas' main tactical goal during this phase is to survive and prepare for the time when they can launch an all-out struggle. This period is characterized by small raids and sabotage, often designed to acquire weapons, stores, and equipment. Equally important at this time is the guerillas' strategic effort to gain the support of the people by impressing them with the gaining momentum of the movement. Every guerilla action is aimed at achieving the desired psychological impact upon the people, the enemy, and the guerillas themselves.

Stage 4. The all-out struggle to swing the balance of power to the resistance movement. The guerillas now feel strong enough to begin widespread activity on a large scale. They may actually gain control over large portions of territory. Their strength and popular support continue to gain momentum. Assistance from outside sponsors may become more obvious. At the same time, the strength of the anti-guerilla forces is drained away.

Stage 5. Evolution into conventional mobile warfare. Guerilla warfare cannot of itself attain complete victory. Guerilla forces seek to weaken enemy power and build up their own strength to the extent that the balance of power will ultimately permit full-scale, conventional actions. Guerilla forces evolve through the earlier stages into regular forces in any area desirable as the situation progresses and permits. Guerillas have a tendency to preserve their better units until they are ready to commit them in this final stage. Usually guerillas must rely on training cadres and equipment from outside sponsors to be able to develop conventional forces. When, through well-planned timing, irregular forces reach this stage of development, their victory is almost inevitable unless outside help on a large scale comes to their enemies.

One can see that the resistance movement has a built-in "snowballing" action, which enables it to gain momentum and grow larger as it rolls along. Conversely, the anti-resistance movement is sapped of its strength as it becomes entangled in the all-encompassing chaos inflicted on it by the resistance. Thus, the anti-resistance force loses the capacity to save itself over time, even though it may possess the military strength to do so. This is no accident—it is the constant task of the guerillas to nurture this action and keep it going.

A critical strategic area for irregular leaders is determining the proper time to put aside guerilla tactics and wage conventional warfare. There is sometimes a strong urge to rush the evolution into regular campaigns. Impatience or hastiness by the guerillas in executing their timetable can result in nasty setbacks. The Viet Minh discovered this, to their near demise, during the earlier stages of their war with the French. This is in violation of the rule that there must never be a lost battle or a skirmish for the irregulars, particularly in the earlier stages. A defeat could result in a forced regression back into one of the earlier stages. It could break the back of the resistance through a loss of morale and popular support. Even so, the resistance may yet be left with enough power to begin the comeback momentum all over again. This is particularly true if anti-resistance forces fail to follow up their opportunities.

There are many relative factors which, if known to the leadership of one side or the other, can be used to increase the alternatives available. The resistance may be strictly an internal affair, or it may be an imported revolution to some degree, but rarely have successful guerilla movements been developed without the support of an outside power.

There have been no historical deviations from these five stages in a resistance movement, although they are sometimes difficult to identify. Mao Tse-tung lists three stages rather than five in his works. His three stages correspond to the last three stages described here. Yet we can see that the trend of events definitely derives itself from the earlier two stages, which we have added in an attempt to clarify the picture.

Also, several of these stages may be developing at the same time at different areas within the conflict zone, giving the impression of an explosive chain of events. The situation will be fluid, to say the least. There will be areas controlled by each side and areas that are contested for but controlled by neither side. The degree of control either side may exercise over a given area may vary with the time of day or night, or with the actual physical occupation of the ground. A hasty analysis of the situation often makes it appear that some of these stages were skipped. Changing circumstances from event to event make it difficult at times to distinguish among the five stages. But they *can* be identified. A sound knowledge of these phases can contribute mightily to a clear analysis of the situation.

One of the greatest boosts to an irregular movement, along with securing popular support, is the ignorance *at the strategic levels* of how the enemy operates in this type of struggle. If the irregulars can progress into the third or fourth stages in developing the conflict *before* their enemies seriously begin to size up the situation, they have won half the battle. The greatest contribution by counter-guerilla forces to their own defeat has been, at the policymaking levels, continual underestimation of the enemy. Their leaders have misunderstood the anatomy of this type of struggle and failed to take the developing war clouds seriously enough until it is too late.

Application of the Principles of War to Irregular Warfare
It is essential to understand the differences in the application of the Principles of War to guerilla versus conventional warfare. Besides failure to understand the anatomy of irregular warfare, one of the greatest faults of leaders has been failure to interpret these differences. This resulted in strategic-level decisions that, especially early in the conflicts, proved dangerous. It resulted in commanders of anti-guerilla conventional forces only using their knowledge of conventional warfare to interpret and react to available intelligence concerning guerilla actions.

Prior to forming a brief analysis of the application of the Principles of War to irregular warfare, we would do well to outline them:

1. Objective	Direct all efforts toward a decisive and obtainable goal.
2. Offensive	Seize, retain, and exploit the initiative.
3. Mobility	Move forces to favor accomplishment of the objective.
4. Mass	Maximum possible dispersion, and adequate concentration at the decisive place and time.
5. Economy of Force	Preserve forces for the decisive blow. Use minimum combat power for secondary efforts.
6. Unity of Command	Assign each effort a single responsible commander.
7. Simplicity	Use simple plans, give concise orders, be flexible.
8. Suprise	Take the enemy unaware.
9. Security	Protect against surprises at all times, in all places. Remain alert.

These principles apply in some form of application to all forms of warfare. Guerilla warfare cannot afford to ignore them. Opponents of guerillas cannot afford to be ignorant of them, particularly at the top levels. For it is at the top that the atmosphere will be set for combating irregulars. It is there that the key decisions of the struggle will be made (or not made) very early in the game, while the balance of power is still favorable. It is amazing how some of the most brilliant manipulators of these principles in mobile warfare have been caught completely off base upon finding themselves embroiled in a revolutionary struggle.

Now let us look at each of the Principles of War in more detail:

The Principle of the Objective. In irregular warfare, the Principle of the Objective (point 1 above) is vitally important in its political sense, to the degree that every last citizen becomes involved. The citizenry has become concerned to such a degree that a significant part of it takes up arms against the established order. The overriding objectives of the struggle, as Mao Tse-tung says, are the aspirations of the people which the resistance movement must translate into effective programs for action. The established order may have the objective of maintaining the status quo, or perhaps a slower program of reform, which does not accelerate rapidly enough to suit the impatience of the revolutionaries.

When these conditions of hyper-political tension are effectively compiled into a meaningful series of objectives in a program of action by the revolutionaries, critical things have happened. First, the revolutionaries have successfully defined and publicized their objectives in a popular manner. These are politicoeconomic, not military objectives. Second, the established order has at this point found itself forced to accept the struggle on these terms, whether it acknowledges it or not. The powers that be must take this factor into consideration when they define and announce their own objectives, if indeed such a process takes place. The strategic levels at this point will find the alternatives available fewer in number and undoubtedly less savory in nature. Once violence has been committed on behalf of the revolution, the government must conjure a positive reaction that meets the objectives of the people in terms they understand. This critical decision is made prior to or very early in the struggle. The cause of the struggle will be political survival and the status quo or, at most, an orderly progression of events toward the general aspirations of the people.

Thus we can see that the nature of the Principle of the Objective in a revolutionary struggle is one of great internal conflict over the economic and political desires of the individual people. The individual people fighting for a stated objective often involve themselves much more intimately, including to the degree of all-out violence, than a nation normally involves itself in war. Because of the politico economic nature of such a struggle, it is vital for anti-guerilla strategists to grasp a popular objective and define it early, clearly, and adequately enough for all to understand.

The Principle of the Offensive. An inherent characteristic of guerillas is offensive action. This is made possible by their application of other Principles of War. It consists of myriad small offensive actions which grow in scope as the guerillas progress from stage to stage in the conflict. Mao Tse-tung has written that offensive action permits a military force to impose its will upon the enemy. The irregulars are able to seize the initiative at will when their opponents allow them to do so. This is a result of the wrong decisions (or no decisions) happening at the strategic levels of anti-guerilla governments before the irregulars possess the power to seize the initiative, even at the smallest scale.

The Principle of the Offensive is not applied in revolutionary wars from a military standpoint alone. The Offensive is a trinity in nature, with psychological, economic, and military approaches all in one. One of the secrets to the Offensive's success in irregular warfare is the application of this trinity concept through great coordination. Its success is made more pronounced through the utter lack of understanding on the part of anti-resistance forces at key moments. The revolutionaries seek to unbalance all three areas of the anti-resistance: a guerilla is a propagandist, a walking economic aid program, and a soldier, all in one.

There are no precedents for anti-resistance forces as to just how the Principle of the Offensive can be approached in guerrilla warfare. Indeed, there is only an awakening going on at the present concerning the existence of this phenomenon. We might suggest that soldiers of the anti-resistance forces seek to reflect this trinity approach, but with emphasis upon areas more suitable to their situation. The psychological warfare and economic warfare fields must be vigorously pressed from the top, with well-coordinated results coming out at the bottom in ways that the people can easily understand. The organization of this type of counter-offensive must be different from how the revolutionaries organize things, but the same trinity idea should apply. Whereas the soldier must be a modified version of the guerilla, his primary effort remains the same: taking up arms. Likewise, the economic warrior might bring assistance to the people in the form of help they can actually understand, but he must be a propagandist as well. Finally, the psychological warfare expert must use all of the media at his disposal to guide the overall psy-war effort of the counter-offensive. Without this trinity approach to the Principle of the Offensive, the anti-resistance forces will remain in the plight similar to a situation recently described in a news report from Vietnam: "The Vietnamese soldiers seem like aliens in their own country."

The Principle of Mobility. One of the vital points here is adapting oneself to the terrain to "get thar fustest with the mostest," as Nathan Bedford Forrest stated. Guerilla warfare is most adaptable to difficult terrain, i.e., jungles, mountains, forests, vast marshes, etc. There, the key to mobility is the foot soldier and pack animal. There, the irregulars with their meager resources can gain more favorable odds. This will contribute to their ability to seize the initiative. The anti-guerilla government which sends a conventionally equipped and oriented army against the irregulars will be cooperating in its own destruction. Such policymakers will probably have made other errors at vital times as well, for their whole thinking will not be broad enough. Lt. Col. Neal G. Grimland in his article "The Formidable Guerilla" (*ARMY*, February 1962) states a logical solution to the mobility problem when he states that conventional units must strip down to essentials and fight as light infantry.

The Principle of Mass. By their mobility and initiative on the ground of their choice, the irregulars will be able to concentrate superior forces and manpower at the right places and times. They will be able to scatter when the actions are over. They will accomplish this over and over, constantly sapping the strength of their adversaries in numerous small engagements. Yet they may face an enemy who may be vastly superior in terms of total manpower and equipment. But they will roll on their momentum of gathering strength. Their adversaries will grow weaker and weaker as they become entangled in a web (partly of their own making), from which they cannot extricate themselves. If high-level policy does not or cannot commit sufficient resources wisely to cope with the situation, the end will become more and more predictable.

But regular forces which *can* be brought to bear in superior numbers against the guerillas will have all the odds if proper tactics are used. The dependence of this principle upon others, including Mobility, Surprise, and Security, is obvious. Airborne, airmobile, or footborne light infantry with air-ground support will be necessary to close the trap. Here, the application of the Principle of Mass implies that the use of adequate forces to surround the guerillas and squeeze them to death is more tactical in nature. From a strategic point of view, we infer that sufficient forces of the right types must be provided for tactical application to take shape.

The Principle of Economy of Force. To the guerillas this means one thing: survive and save the weapons. This is why guerillas thrive on hit-and-run tactics. In the earlier stages of the conflict, this is an over-ruling factor. As the guerillas manage to develop some better-trained and -equipped units, they will show a tendency to preserve these units if necessary, at the expense of poorer-quality units. This can be illustrated by the policy of the Viet Minh against the French in Indochina. The Viet Minh used poorer-trained and -equipped local guerilla units to fight delaying and screening actions against the French. The purpose of these actions was to enable better-trained and -equipped Viet Minh units to escape a decisive engagement when the balance of power was not deemed in their favor. *(This gave them the ability to refit in Laos and Cambodia.)*

For anti-resistance forces, the meaning of Economy of Force includes two important factors. These factors have long-range implications, and this in itself makes them prone to be taken lightly by human nature. Basically they mean, "Do not put off until tomorrow what you can do today." Prompt and positive action (or at least reaction) will save later losses or the necessity to commit far greater resources later. Economy of Force is related intimately to time factors in revolutionary warfare. The two factors we have suggested are:

1. To regain the initiative at the earliest opportunity to deny the irregulars freedom of action to snipe away at our valuable resources.
2. Destroy as many guerillas as soon as possible and seize their resources from them. Deny them access to resources, theirs or ours.

The connotation of Economy of Force here is different from the conventional application for anti-guerilla forces in the sense that it means to do today to prevent greater loss or even disaster tomorrow. It has an offensive-defensive sense in that it seeks to turn the initiative and momentum away from the guerillas. This will of course assist in attaining the final objectives of political victory. But it will also prevent anti-resistance forces from being worn down in the evergrowing, ever-entangling net of the revolutionaries, to the point where the balance of power or the will to fight is lost. This cannot be accomplished tactically without a strategy that foresees these things.

The Principle of Unity of Command. Compared to some of the other principles, there is less difference in application of the spirit of Unity of Command to guerilla versus conventional forces. Violation of this tenet would be equally disastrous to either side. When revolutionaries violate this principle, they face defeat in detail. Violation by anti-irregular forces may not bring such immediate and spectacular results, but lack of information, and uncoordinated and ineffective action become a weakness made to order for the irregulars to exploit.

The Principle of Simplicity. This has two important meanings in guerilla warfare. To anti-guerilla conventional forces, there is the danger of interpreting it in the conventional-warfare sense. To guerillas it means working successfully with fighting men, many of whom may be peasants lacking in tactical sophistication. This principle calls for the simplest of weapons and organization. Conventional types of anti-resistance forces have made the mistake of looking down their noses with disdain at the simplicity of the guerillas. It is astonishing that some have continued to do so, even after ultimate defeat at the hands of irregular forces. It is important for conventional forces opposing guerillas to:

1. Be alert and flexible enough to be able to develop the right mental approach toward this mode of warfare, and
2. Reorganize and simplify to adjust to the situation imposed upon them by this mode of warfare.

Top-level policy must nurture this attitude in the military early enough to prevent undue losses. This means that the use of modern mechanized equipment and some aspects of airpower must be seriously reevaluated in terms of the situation. Traditional concepts must give way in planning circles to realistic, flexible thinking. The mechanized technology of the Western military mind must be ready to switch to the simple technology of the light-infantry foot soldier. We must see that our desire for rapid shock action may best be realized by the airborne, airmobile, and foot soldier in guerilla warfare. All of the unnecessary, cumbersome paraphernalia so vital in conventional warfare must be trimmed down as soon as it becomes obvious its use is too disadvantageous. Conversely, the very simplicity imposed upon most guerilla movements because of their lack of resources and training can be turned against them. They cannot stand against properly equipped, well-trained regulars, should these troops be led into a position where a decision can be forced. Proper guerilla warfare application of the Principle of Simplicity by anti-resistance forces will enable the proper application of the other principles, as all these things bear upon one another.

The Principle of Surprise. The essence of irregular warfare, with the initiative firmly planted in the hands of the guerillas, leads to an increasing number of small surprise attacks. This type of small-unit action, quick hitting with a rapid withdrawal, is essential to the survival of the guerillas until they progress into a stage whereby they can wage sustained combat. Conversely, surprise actions by anti-guerilla forces under these conditions are extremely difficult due to the guerillas' extensive security system. Yet surprise actions by anti-guerilla forces are vital if the irregulars are to be caught and brought to bay. It is the responsibility of the strategic levels of anti-guerilla governments to provide conditions which will enable the military to launch surprise blows upon the resistance. This may mean such extreme measures as isolating the guerillas from their main source of information, namely the people. The counterintelligence and security apparatus necessary to protect surprise operations becomes more difficult to provide for the anti-guerillas with time. Obviously the best time to launch all-out offensive operations against guerillas is in the early stages of the movement, before a significant proportion of the people commit themselves to the resistance. Provisions for secrecy must go far beyond those of the past. Decoy operations, fakes, and feints of all descriptions must be carried out at all levels. Troops involved may be isolated prior to operations and possibly told little or nothing until potential compromise of information is no longer an important factor.

The importance of adequate intelligence and counterintelligence networks can be appreciated. Resistance movements which depend so much upon surprise and security generally have the most adept intelligence and counterintelligence apparatus. Anti-resistance movements, on the other hand, have been generally weak in these fields, and understandably so, if the people favor the resistance. Even so, potential anti-resistance governments that recognize the possibilities of their position could go much further in prior preparations in the intelligence fields. We mention the intelligence fields here because surprise depends on information and on denying the enemy information, and these have been key factors in guerilla warfare. Guerillas, to be destroyed, must be caught by surprise. Guerillas themselves thrive on surprise actions. All this depends upon acquiring and denying information.

The Principle of Security. Both sides in a revolutionary war are vitally concerned with security in all its aspects. The guerillas are more concerned with counterintelligence, early warning, and evasive action than with fortified positions. It is the inaccessibility of guerilla bases which provides much of their security. The anti-guerillas have usually been in a predominantly defensive posture and found themselves so bogged down, with resources tied up, that the majority of effort involved passive defensive measures to protect possible guerilla targets. But the best security for anti-resistance forces is in the form of vigorous offensive campaigns which restrict the guerillas' freedom to strike at target areas. Unless the strategic levels plan for these operations early enough, things will get out of hand, and resistance strategy will force passive defensive reactions. Thus resources may become spread so thin in the attrition of constant defensive actions and half-measures that the giant may die under a swarm of mosquitoes.

The very nature of guerilla warfare requires both sides to use all forms of active and passive security measures with a super-sensitive alertness.

Conclusion

What we have attempted to do here is underline the obvious: that strategic-level decisions (or lack of them) at top governmental and military levels are where revolutionary-type wars are won or lost. The most brilliant tactical-level operations stand a poor chance without knowledgeable, competent national leadership in areas relating to the subject.

By this we mean not only in the military areas but especially in the psychopolitical and economic areas. This is not meant to detract from the importance of tactical-level competence, only to say that strategy is vital. Whichever side possesses superior strategic-level leadership will most probably produce a highly competent tactical plan as well.

SOURCES

Chandler, David G. *The Military Maxims of Napoleon* (1988).

Clinton, James W. *The Loyal Opposition Americans in North Vietnam, 1965–1972* (1995).

Ezell, Edward C. *Personal Firepower: The Illustrated History of the Vietnam War* (1988).

Fall, Bernard B. *Street Without Joy* (1961).

Fallows, James. "What Did You Do in the Class War, Daddy?," *Washington Monthly* (1975).

Federer, William J. *America's God and Country Encyclopedia of Quotations* (1994).

Gabriel, Richard A., and Karen S. Metz. *From Sumer to Rome: The Military Capabilities of Ancient Armies* (1991).

Halberstam, David. *The Best and the Brightest* (1972).

Hamilton-Merritt, Jane. *Tragic Mountains: The Hmong, the Americans, and the Secret Wars for Laos, 1942–1992* (1992).

Hendrickson, Paul. *The Living and the Dead: Robert McNamara and Five Lives of a Lost War* (1996).

Katz Keating, Susan. "Plausible Denial," *Air & Space* (May 1997).

Krepinevich, Andrew F., Jr. *The Army and Vietnam* (1986).

Ky, Nguyen Cao. *Twenty Years and Twenty Days: How and Why the United States Lost Its First War with China and the Soviet Union* (1976).

Leverington, Karen, ed. *The Vital Guide to Fighting Aircraft of World War II* (1995).

Lunev, Stanislav, with Ira Winkler. *Through the Eyes of the Enemy: Russia's Highest Ranking Military Defector Reveals Why Russia Is More Dangerous Than Ever* (1998).

McMaster, H. R. *Dereliction of Duty: Johnson, McNamara, the Joint Chiefs of Staff, and the Lies That Led to Vietnam* (1997).

McNamara, Robert S. *In Retrospect: The Tragedy and Lessons of Vietnam* (1995).

McNamara, Robert S. *Argument Without End: In Search of Answers to the Vietnam Tragedy* (1999).

Means, Howard. *Colin Powell: Soldier/Statesman, Statesman/Soldier* (1992).

Newman, John M. *JFK and Vietnam: Deception, Intrigue, and the Struggle for Power* (1992).

Olson, James S., and Randy Roberts, *Where the Domino Fell: America and Vietnam, 1945–1990* (1991).

Osanka, Franklin Mark, ed. *Modern Guerrilla Warfare* (1962).

Palmer, Bruce, Jr. *The 25-Year War: America's Military Role in Vietnam* (1984).

Powers, Richard Gid. *Not Without Honor: The History of American Anticommunism* (1995).

Prochnau, William. *Once Upon a Distant War: David Halberstam, Neil Sheehan, Peter Arnett—Young War Correspondents and Their Early Vietnam Battles* (1996).

Prouty, L. Fletcher. *JFK: The CIA, Vietnam and the Plot to Assassinate John F. Kennedy* (1992).

Record, Jeffrey. *The Wrong War: Why We Lost in Vietnam* (1998).

Rust, William J. *Kennedy in Vietnam: American Vietnam Policy, 1960–1963* (1985).

Schanche, Don A. "Last Chance for Vietnam," *Saturday Evening Post*, January 6, 1962.

Shaplen, Robert. *The Lost Revolution: The U.S. in Vietnam, 1946–1966* (1965).

Smith, Patricia. "Doctor Smith and 300,000 Tribesmen," *The Sign*, April 1962.

Summers, Harry G., Jr. *Historical Atlas of the Vietnam War* (1995).

Summers, Harry G., Jr. *On Strategy: A Critical Analysis of the Vietnam War* (1982).

Summers, Harry G., Jr. *Vietnam War Almanac* (1985).

Webster's New World Dictionary of the Vietnam War (1999).

Winters, Francis X. *The Year of the Hare: America in Vietnam, January 25, 1963–February 15, 1964* (1997).